The Form *of* Time

Books by Elliott Jaques

•

Changing Culture *of* a Factory

Measurement *of* Responsibility

Equitable Payment

Product Analysis Pricing
(*with Wilfred Brown*)

Time-Span Handbook

Glacier Project Papers
(*with Wilfred Brown*)

Progression Handbook

Work, Creativity *and* Social Justice

A General Theory *of* Bureaucracy

Levels *of* Abstraction *and* Logic *in* Human Action (*Editor*)

Health Services (*Editor*)

The Form of Time

Elliott Jaques

Crane Russak • *New York*
Heinemann • *London*

The Form of Time

Published in the United States by

Crane, Russak & Company, Inc.
3 East 44th Street
New York, New York 10017
ISBN 0-8448-1394-x

Published in Great Britain by
Heinemann Educational Books
22 Bedford Square
London, WC1B 3HH, England
ISBN 0-435-82480-5

Library of Congress Cataloging in Publication Data

Jaques, Elliott.
The form of time.

Bibliography: p.
Includes index.
1. Time. I. Title.
BD638.J36 115 81-17510
ISBN 0-8448-1394-X AACR2

Printed in the United States of America

Contents

To Kate

Acknowledgments

I have been preoccupied for the past thirty years or more with the idea of time as fundamental for measurement in the human sciences. More recently my concern has been to generalize that idea by exploring the nature of time itself. I have been helped by many discussions with John Isaac, Roland Gibson, and Gillian Stamp, as we struggled with ideas which have been reported in our book *Levels of Abstraction in Logic and Human Action;* and by separate discussions with William Helme, who has also commented on a draft of the manuscript. My discussions with these colleagues have been, and continue to be, a source of intellectual stimulation and pleasure, and I am grateful to them.

The final chapters of this book were written in the tranquillity of the Center for the Humanities at Wesleyan University.

Rhoda Fowler has now prepared manuscripts for me for over twenty-five years, and I gratefully acknowledge her help in editing, in typing, and in preparation of indexes. She has exercised her usual care in the preparation of this book.

The extracts from ''Burnt Norton'' from *Four Quartets* by T. S. Eliot are reprinted with the kind permission of Faber and Faber Ltd. of London and Harcourt Brace Jovanovich Inc. of New York.

THEMA

These are the forms of time, which imitates eternity and revolves according to a law of number.

Plato, *Timaeus*

. . . it occurred to me suddenly that, if I had indeed the strength to accomplish my work, this afternoon—like certain days long ago at Combray which had influenced me—which in its brief compass had given me both the idea of my work and the fear of being unable to bring it to fruition, would certainly impress upon it that form of which as a child I had had a presentiment in the church at Combray but which ordinarily, throughout our lives, is invisible to us: the form of Time.

Marcel Proust, *Time Regained*

A Summary of the Main Argument

The nature of time has been a central and continuous subject of philosophical controversy for over two thousand years. There are many unresolved questions. Does time flow? Is there an arrow of time and does it fly in one direction? Is the future different from the past? Is time simply the perception of motion? Can equal spans of time be measured? Is there such a thing as time at all?

It is the argument of this book that the failure to alleviate the confusion bound up in such questions and to establish a sound philosophical understanding of the nature of time—especially time as abstracted from and integrated back into the space–time manifold—has been a major factor in the lack of development of a sound understanding of the nature of man and of society. By contrast, the relatively easier task of constructing a sufficient philosophical base for the conceptualization of space and of things—in essence the Euclidean and Cartesian systems—gave the necessary foundation for the enormous growth and power of the natural sciences. The clarification of the time dimension(s) of the space–time manifold is an essential condition for putting the social sciences on an equally rigorous foundation.

The grounds exist for resolving these and similar questions as a result of a significant convergence in thought upon two different meanings of time: one meaning relates time to the experience of purposive, intentional, goal-directed behavior; the other meaning relates it to the sense of passing time expressed in successive readings of a clock. This convergence points to a conception of time as *two-dimensional* and not, as is most commonly assumed in relativity as well as in philosophy, as unidimensional. It further suggests that direction is a psychological phenomenon associated with intention and not with flow.

The book begins with the problem of defining time in the sense of establishing what is included under this concept. Common parlance suggests a multiplicity of different kinds of time. In fact, there is only one kind of time—the continuity of existence of things. It is this assumption of continuity which is necessary to describe events, process, or change; that is to say, to define an event, or process, or change requires the conception of the same things continuing to exist, but continuously modifying their position, or condition, or both.

This notion of time as the experience of continuity can then be shown to be the same conception of time, whether experienced subjectively or objectively, in working or sleeping experience, in cyclical or linear processes, or in any of the many other usages.

Time having been defined, the experience of time is then considered. It is here that the two-dimensionality of time emerges. The first dimension is the temporal axis of intention, the time by which we intend, or plan, or seek to accomplish something; and the second dimension is the temporal axis of succession, the axis along which we standardize our calendars and clocks.

This two-dimensional analysis is then shown to solve one of the great problems of time: how to deal with the ideas of past, present, and future, of earlier and later, of before and after, of passage and direction, of flux and constancy, of change and permanence, of durée and atomism, of continuity and discontinuity.

Our ideas of past, present, and future, passage and direction, flux and change, durée and continuity, are exclusively associated with the temporal axis of intention. They are expressions of our field of experience in the flowing present. The past is the experience of the flow of memory, the present of perception, and the future of expectation and desire. Durée, flux, and passage are the experience of the interaction of memory, perception, and desire, and direction is the experience of the goal-directedness of intention.

Our ideas of earlier and later, before and after, temporal discontinuity and atomism, constancy and permanence are, by contrast, exclusively associated with the temporal axis of succession. They are expressions of our experience of a cross-sectional or spatial abstraction from events, expressions of our capacity mentally to make time for the moment stand still, while we record as in a photograph the happening of that moment. We can then date this frozen moment on a calendar or clock, as having occurred on such and such a day at such and such a time.

It is at this point that the argument takes up the usefulness of having the duality of both a continuous field conception and an atomic or discrete object conception, with cognitive oscillation between the two. In terms of time, it takes us to the notion of continual cognitive oscillation between field and points at a distance, between intention and succession, between past, present, and future and earlier and later, between flux, change and continuity and permanence, constancy and discontinuous beads of time.

From this point on the book turns to the particular significance of starting with time rather than with space, in considering the living as against the material world, the world of psychological and social processes and events as against the world of physical things (whether still or in motion). Purpose, intention, goal-directedness are seen as the essential qualities of all living beings including human beings as agents engaged in self-initiated working activity, alternating with reflection, sleep, and dreaming.

The unconscious and preconscious mental organization of processes in time are then seen as giving the substratum or feel of time, the sense of time, intuitively felt as flow or durée. These unconscious and preconscious processes are experienced as fields in which memory, perception, and desire, and past, present, and future interlace and interact in feeling without boundaries between, as a whole giving the background against which our conscious awareness of time stands out.

The conscious mental awareness of time, by contrast, is that of figure against ground, of focused awareness of points in time, some earlier and some later. And as is the case with all focused perceptions of figure, these points in time are bounded, verbalizable, recordable, and material for objective scientific measurement and control.

Out of these formulations there then emerges the notion that man's life is lived in intentionally directed and organized episodes or events. These intentions will be more or less unconscious and more or less conscious, the two interacting as ground and figure. Moreover, the intentions are directed toward goals which have a planned time of realization as an integral part of their essence. This fact allows for the recognition of the time-frame of the individual: the frame, that is, that bounds the farthest forward in time of the goals which an individual is capable both of constructing *and* of achieving. This time-frame turns out to be an objectively measurable characteristic of each individual; and the argument is then pursued that the significance of this time-frame is that its size measured in time scale gives a direct measure of the size of the psychological capability of the individual, of the temporal domain he occupies, of the size of the world in which he lives and which he is able to organize and cope with, of the size, in short, of the life space he inhabits.

Further evidence is then presented to show that the size of the time-frame of individuals matures and develops in a systematic and predictable manner throughout life—from seconds in infancy to days, weeks, months, and years in adulthood, the actual size of the fully-developed time-frame depending upon the capability of the individual. It is at this point in the argument that the possible significance becomes apparent of taking intentional events and episodes, and their duration measured in time, as the starting point and foundation for the psychological and social sciences.

It is further argued from these temporal measurements of the size of the time-frame that if in the social sciences we start with measurement in a 2-dimensional time context and a 5-D world view, it becomes possible to move directly to equal-ratio-scale measurement of the same objective type as that (starting with the measurement of length) which is the major instrument of the quantitative natural sciences. This step avoids the pretence of the spurious ''hard numbers—hard science'' game currently played with what are really very soft-headed spatialized ordinal-scale ratings that masquerade as objective scientific measurement in the psychological and social sciences today.

The final section of the book reviews all these matters in a more general context. The nature of psychological and social things as temporal things is examined as an epistemological issue. And lastly, the nature of cognitive oscillation between the field perspective and the atomic perspective in understanding the world is analyzed in more detail, and their significance for various views of the world—mechanistic, materialistic, mystical, and holistic—is considered in relation to science and to everyday life.

The theme of these final chapters is that even in the natural sciences an exclusively atomic view of a world of constant things in space is now not only untenable as a dominant philosophical world view, but is in fact harmful. This atomic view, which had great value in the development of the physical sciences and technology, still has value as a philosophical basis for those aspects of human activity concerned with what might in general terms be called the mechanical. But it needs to be complemented by a field theoretical view of the world, and interrelated with it, for a comprehensive and nonfragmented understanding to emerge.

The conclusion that field theory is needed as well as atomic theory has long been established. What has been less clearly noted, however, is the consequence of this conclusion for the conceptualization of time. The discontinuous view of time as points along a string which accompanies the atomic outlook must be complemented (and not supplanted) by the Heraclitean continuous flux which is the time most relevant to force fields.

The philosophical outlook stemming from the Heraclitean view of a world in flux requires proper recognition, so that a living conception of flux along with discrete objects-at-a-distance may come into the center of human thought. This complex view in terms both of atoms and of flux is essential for the development of the life sciences—of living processes and activities in biology, psychology, sociology, and other social sciences.

What is needed, then, is not a switch from the atomism of Parmenides to the flux of Heraclitus. It is in fact a duality—a genuine dilemma—which is required, one in which the space–time manifold becomes the center-focus of reality, with recognition of the rapid cognitive oscillation between two different sets of constructs of space and of time seen as abstractions from the manifold. As far as time is concerned, this duality with oscillation between the poles allows, on the one hand, for a temporal time—that is, a time with temporal extension and with content; as well as allowing, on the other hand, for a spatialized time—that is, a time which is made to stand still while objectified things are dealt with as though they were for the time being still and unchanging.

It will be demonstrated that to locate and study social and psychological things requires in the first place that they be abstracted from the space–time manifold as events or episodes extended in flux–time, just as to locate physical things requires in the first place that they be abstracted as entities extended in space. In the fortunate language of Stuart Hampshire, these events or episodes are described as trajectories—a term which I shall borrow.[1] In short, it will be argued that the essence of social and psychological "things" is their extension in time as flux, and their study requires the study of the form of their trajectory through time.

[1] Stuart Hampshire, (1965), *Thought and Action*.

PART ONE

PHILOSOPHY OF TIME AND ACTION

CHAPTER ONE

The Enigma of Time

The enigma of time is the enigma of life: it has plagued poets and philosophers from the beginnings of civilized thought. For life is lived in time. Without time there is no life. But each one lives in his own time. No two men living *at* the same time live *in* the same time. Each one, living at the same moment, has his own personal time perspective, his own living linkage with past and future, the content of which, and the scale of which, are as different between one person and another as are their appearance, their fingerprints, their characters, their desires, their very being.

That any two persons should differ in the contents of their thoughts, desires, memories, aspirations is self-evident. People are different. That gives richness to life. Indeed, the freedom for individuals to be different, to behave differently and in entirely new ways, to think differently, to be idiosyncratic, within reasonable boundaries is the essence of freedom, of the liberty of the person, of individual normality, of the requisite society, and of the quest for life.

But that different people live in different time scales, or in different *temporal domains* as I shall refer to them, may not be so self-evident. Yet it has profound and far-reaching consequences for everyone. It is through the recognition of these different time scales within which people live that many of the mysteries of time can be resolved, and time itself may be understood.

The ordinary conception of time as used in clocks and in the natural sciences will not, however, be sufficient for our purposes. Even the time of relativity theory, the time which intermixed with (rather than merely added to) the three dimensions of space gives the modern 4-dimensional dynamic view of the world—the physical world, that is—will not by itself satisfy the requirements of the life sciences. For the time of relativity theory is still the same old clock time used for dating simultaneous points at a distance in physical events. It does not live or breathe.

The things which make for life, which make life different from physics, require for their description a sense of time which encompasses memories in the present of the past as well as expectations and desires in the present of the future. Living time extends into feelings for the past and desires for the future which no physical object can possibly experience.

To keep our descriptions of living things alive—both individual and social processes—requires not the cross-sectional spatial abstraction which kills life dead, but a longitudinal temporal abstraction of living events. Thus, our study has as its prime purpose to understand the form of time, the nature of time, the form which living processes take in the course of their movement in time. To understand the nature of time is thus to enhance our understanding of ourselves, of our actions and our social relationships, and of human life itself.

Past, Present, and Future

All the difficulties in understanding the meaning of time are contained in the riddle of past, present, and future. Are they co-terminous? Does one flow into the other? Does the future become the present and then the past? It is these questions which were raised centuries ago by St. Augustine, and later in remarkably similar terms by T.S. Eliot, by William Faulkner, by Proust, by Dos Passos, by Joyce, by Virginia Woolf, all of whom have attempted to reflect the experience of time in their writing.

St. Augustine stated the mystery in words which ring and echo down through the ages: "For what is time? . . .. What in discourse do we mention more familiarly and knowingly, than time? And, we understand, when we speak of it; we comprehend also, when we hear it spoken of by another." And then the famous dictum: "What, then, is time? If no-one asks me, I know: if I wish to explain it to one that asketh, I know not; yet I say boldly that I know, that if nothing passed away, time past were not; and if nothing were coming, a time to come were not; and if nothing were, time present were not." [1]

But St. Augustine did not let matters rest with mere questions to which there were no apparent answers. With his evergreen thought, his two-thousand-year-old modernity, his psychological insights ahead of his time and his setting, leaping across the ages yet to come, and entirely out of keeping with the state of development of the human intellect and human understanding of his own time, he saw the dilemma, the enigma, and he pointed to a significant element for its solution. It was as though he were solving the riddle of the Sphinx, recognizing that time past, time

[1] St. Augustine, *Confessions,* Book XI, Section 14, p. 261.

present, and time future exist not just in the mind of man but as the essence of the mind of man, in the form of the interaction of memory, perception, and anticipation or desire, which enables each one to pursue his life's aims. In the following single paragraph (which is far less often quoted than his dramatic proclamation of the elusiveness of time), he outlines the psychological meaning of time, and does so in modern terms.[2]

"From what we have said it is abundantly clear that neither the future nor the past exist, and . . . it is not strictly correct to say that there are three times, a present of past things, a present of present things, and a present of future things. Some such different times do exist in the mind, but nowhere else that I can see. The present of past things is the memory; the present of present things is direct perception; and the present of future things is expectation. If we speak in these terms, I can see three times and I admit that they do exist".[3]

Later, St. Augustine drives home his psychological analysis: "It can only be that the mind, which regulates this process, performs three functions, those of expectation, attention and memory. The future, which it expects, passes through the present, to which it attends, into the past, which it remembers."[4]

This recognition of the possible coexistence and conjunction of time present, time future, and time past, is a recurring theme in T.S. Eliot, in whom the Augustinian influence shines through, achieving its fullest expression in *Four Quartets:*

> Time present and time past
> Are both perhaps in time future
> And time future contained in time past.
> . . . .
> Time past and time future
> What might have been and what has been
> Point to one end, which is always present.
> . . . .
> And the end and the beginning were always there
> Before the beginning and after the end.
> And all is always now . . ..[5]

[2] J.J.C. Smart, for example, and Richard Gale in their admirable essays on time both omit this paragraph from their edited versions of St. Augustine's *Confessions*. See Smart, (1964), "Questions About Time" in J.J.C. Smart, (Ed.), *Problems of Space and Time:* New York: Macmillan; and Gale, (1968) "Some Questions About Time" in Richard M. Gale, (Ed.), *The Philosophy of Time:* London, Macmillan.

[3] St. Augustine, *Confessions, op cit,* Book XI, Section 20, p. 269.

[4] Ibid., Section 28, p. 277.

[5] T.S. Eliot (1944), "Burnt Norton," in *Four Quartets,* p. 7.

Here, Eliot sets out the same feeling as St. Augustine about the immediacy of time, the all-at-onceness of the past, the present, the future, fused in the ongoing living of human experience as it is happening, as one's feelings occur and ambitions and desires are pursued, as one is hurt, or loving, or seeking, or creating, or working, or performing, bringing something to fruition or beginning to feel the pangs of frustration or of failure—all these experiences of living, of striving to achieve in the future by mobilizing one's past experience and one's present abilities in a composite of activity in which past, present, and future blend in perfect fusion to make up the field of one's real and current flow of psychological existence.

Proust achieved this dramatic effect by the detailed accumulation of events from the past, working out the way in which they remain alive and enter into the present through the minds of each individual actor in the drama. The unfolding of each person's living past and desires enters into the web of interweaving human relationships that unfolds in lace-like complexity.

In the novels of William Faulkner also, the idea of the coexistence of past, present, and future in the human mind is put dramatically to work. Jean Pouillon describes Faulkner's decomposition of chronology as showing "that the present is submerged in the past, that what is lived in the present is what was lived in the past. In this case, the past is not so much an evocation as it is a constant pressure upon the present, the pressure of what has been on what is . . .. The past is not a temporal past, that which no longer is and can only be remembered. It is something here and now, present in the proper sense of the word. Inserted into time, the past *was* and is therefore past, but inasmuch as it subsists, it is present." [6]

There is in Faulkner the disorganizing impact upon the reader of the way in which past and present flow back and forth into each other, all seemingly contemporaneous, so that the present chronology takes on the sense of predestined future when seen through the perspective of the comingled past. It is this comingling of the past with the present and with the predestined future which heightens the vividness and poignancy of the tragedy and grief which hang so heavily over the characters in his novels. Everything seems ominously fated—the future being tied to the accumulating active past.

Breaking away from the predestinating past calls for a breaking away from a static temporality. Thus, Jean-Paul Sartre, writing about Faulkner's *The Sound and the Fury,* places special emphasis upon the symbolic

[6] Jean Pouillon, (1966), "Time and Destiny in Faulkner," p. 68.

import of the breaking of his watch by Quentin, one of the main characters. Quoting Faulkner, ". . . time is dead as long as it is being checked off by little wheels; only when the clock stops does time come to life," Sartre adds, "Thus, Quentin's gesture of breaking his watch . . . gives us access to a time without clocks." [7]

This distinction between clocktime as lifeless time and the fused past, present (and future) as human or living time, is a crucial distinction. Failure clearly to recognize the distinction between these two aspects of time, and to formulate it, and to treat and to sustain the two aspects both separately and in relation to each other—rather than merely the one or the other on its own—has, I believe, seriously impeded the growth of our understanding of time. Another way of formulating the distinction is to separate time as flux (the fusion of past, present, and future) and time as a chronological series of points on a string (the face of a clock). This alternative way of stating the issue is familiar in certain philosophical considerations of time, which I will briefly consider in a preliminary way.

One View of the Philosophical Dilemma

Consideration of the nature of time pervades the literature of philosophy. Inextricably associated with space, it lies at the heart of the search for a coherent picture of the world—a picture which can do justice to the requirements of science, art and culture, religious belief, stringent contemplation and reflection, and the ordinary common sense of everyday life. To know and to understand the main currents and manifold strands of this massive accumulation of thought about the nature of the world in which we live and of which we are a part, is a daunting task for the professional philosopher steeped in the lore and knowledge in his field. It is folly for the interested amateur to undertake any such task.

It would be equally foolhardy, however, to approach the question of the form of time—even with the limited objective of gaining insight into the pervasive framework of human activity and endeavor—without trying to cull as much as possible of the wisdom accruing from the concern of the philosophers. That wisdom confronts the student with a bewildering range of disparate, conflicting, and endlessly fascinating ideas. What I propose to do is to select certain themes as background for the argument I intend to pursue. Among these, the necessary starting point must be the controversy between the view of a discontinuous world made up of fixed and constant entities in empty space, and a continuous world of fluxion, of the insubstantial, of changing patterns of fields of force.

For our present purpose, the argument may be seen as stemming—

[7] Jean-Paul Sartre, (1966) "On 'The Sound and the Fury': Time in the Work of Faulkner," p. 88.

in Western thought at least—from the preoccupation of Parmenides with Being as it is, of Democritus and Leucippus with everlasting atoms and the space they occupied, and of Plato with eternal forms; this will be in contrast to the view expressed in the few fragments left from Heraclitus, expounding the uncertainty of flux.

The sources of these differences are manifold. One influence, however, emanates from the deep-lying human feelings about change and about stability. For Parmenides things *are;* Being *is* and will remain so; all can be known and all can be certain. For Heraclitus all is flowing, changing, transforming, never still, opposite becoming opposite, the only reality being the reality of Becoming itself. In the world of Parmenides, Being can never lead to Becoming, and time is denied any independent reality. In the world of Heraclitus, Becoming can never beget Being, and material substance is no more tangible than—to use his own metaphor—the flickering flames of the fire.

Through most of history since, the secure feelings of certainty which accompany the substantial view have held at bay the anxiety-provoking uncertainty of the continuous-flux description which is as difficult to grasp mentally as it is impossible to grasp physically. In particular, the differentiation by Leucippus and by Democritus of the permanent Being of Parmenides, into a complex of atoms of different shapes and sizes, all located in space and as indestructible as the common substance from which they derived, gave the intellectual starting point for the development of science. And Plato in the *Timaeus* added his powerful voice to this outlook with his argument that the pattern of law is expressed in the form of ideal geometrical shapes which are absolutely at rest and timeless.

The truth is that great successes were achieved when things were somehow made to stand still long enough to be counted and measured. The idea—the abstraction, the fiction—of the absolute constancy of physical things (most physical things, that is) became the necessary foundation of the development of scientific and mathematical abstractions: reflected in the beauty and completeness of the Euclidean system of axioms and theorems, in the discovery of specific gravities and densities, and even in the technique for stopping time by recording it on notches on a stick or on fixed blocks of space in a calendar.

By comparison with the constant spatial orientation which became the essential underpinning for scientific development, the problem of time, of flux, was inordinately difficult to handle. It was of great philosophical, religious, and indeed practical interest, but was not available to the senses in the same direct way as were physical objects, for manipulation and study. The problem of all scientific development from then on was established: physical objects could be systematically studied and measured; time could not be, because it suffered from an apparently insurmountable problem: it would not stay still long enough to be observed, counted, and categorized: its very essence was that it would not stand still at all.

The flux of time which makes it so difficult to get hold of, to hold down, is partly what makes it so difficult to describe. In effect, to put time into words is to do what words always do—they make things stand still. Time is dynamic, words are static; words seem useful for pinning things down, but they ruin time precisely by pinning it down. As Hegel put it, "words murder time." There is in fact no great difference between words and things and words and time, but what was less recognized at the beginning was that the unchanging solidity of atoms, giving the world of unchanging certainty [8] so emotionally acceptable to Leucippus and Democritus and their philosophical followers through history, brought with it the problem of space within which independent and bounded things existed. The things were easy enough to distinguish and the atoms to hypothesize—but the nature of space presented greater difficulties and has continued to do so ever since. "What is space?" has presented no easier a question, no less enigmatic, no less mysterious, than "What is time?"

Nevertheless, the atomic model has proved a useful abstraction. It allows for the sense of certainty of knowing *where* something is and how big it is—and therefore for apparently unequivocal counting in research and economic trade. And it allows for the certainty of knowing enough about what something is, how it is bounded, to make it easy in research and in economic trade to analyze some of its properties in combination with other things.

Motion in an Atomic World

Time, however, does not appear to figure largely in these pursuits. It was not at first necessary for the science of statics, which was growing, other than the gross times needed for astronomy and navigation. The development of dating and of time-of-day calculation which was of interest from our point of view, was the dramatic and disturbing introduction by Galileo of the study of velocity and acceleration—studies made possible by the discovery of the pendulum and of the clock trip mechanism in the West, in the fourteenth century.[9] It is hard now to realize how difficult

[8] That this picture of the world in terms of certainty is the reflection of particular levels and types of human capacity, is argued in Jaques, Gibson and Isaac, (1978), *Levels of Abstraction in Logic and Human Action*.

[9] It was discovered in the eighth century in China, but did not become available at that time in the Western world.

it was at the time to grasp these ideas. It meant adding the uncertainty of the feeling of a rate of motion, and the idea of measuring it, to the certainty of the objects which moved. And not only of velocity, but of directly unobservable and difficult-to-grasp rates of increase or decrease in motion. The dilemma was talking about a rate of acceleration, for example, without being able to stop the object and observe the acceleration at firsthand.

Even so, we may note that time as construed for these purposes was the time of Quentin's watch. It is the time of the trip mechanism; that is to say, the time measured by stopping motion, or in effect, stopping time every second by a cogwheel.[10] It was this intermittently stopped clock time which became the time that was absorbed into the scientific revolution of Galileo's formulation of the inertia of physical bodies, and of Newton's mechanics.

This stopped clock notion of time is of great interest, since in a curious way it is static rather than dynamic; if, that is, it is possible to think of time as static. What I mean is that it is a time that goes along with the atomic view of the world: a time that causes itself to be defined in terms of the apparent motion of an object from one point in space to another, in a given period or lapse. Perhaps a better epithet is that of mechanical rather than static time, of time that brings dynamics to a mechanical world.

It might have been expected that with the emergence of relativity theory, this mechanical time would be replaced by a flux time. Oddly, however, this expectation was not realized. The addition of time as a fourth dimension simply adds to the atomic three-dimensional world of objects. Time as a fourth dimension remains stubbornly out of phase with the isomorphic interchangeability of the three Cartesian spatial coordinates. The problems of simultaneity and succession, even at the speed of light, are resolved—up to a point—but leave in their wake problems of the most serious kind, among them the apparent death of time itself.

The world view of the special theory of relativity leads to a curious four-dimensional block universe in which all time is present at one dead and unchanging time, once and for all, in which all is determined, all is fixed, all is in its place. Each person simply moves on his particular path along his particular part of this creation, like a traveler in a train, or on a traveling staircase or an escalator—a spectator rather than a willful and active participant, watching life go by against the background of scenery

[10] It was an impressively bold and imaginative stroke which was involved in realizing that in order to measure time it was necessary literally to introduce a regulated and recordable *stoppage* of a process.

as it moves along, all unchanging, arranged, predetermined, meaningless, inhuman, and lifeless.

Whitrow described this block universe: "Even Einstein, who made the greatest contribution since the seventeenth century to the understanding of time . . . later became decidedly wary of the concept, . . . and came to the conclusion that physical reality should be regarded as a four-dimensional existence rather than as an *evolution* of a three-dimensional existence. In other words, the passage of time is to be regarded as merely a feature of our consciousness that has no objective physical significance. This sophisticated hypothesis makes the concept of time completely subordinate to that of space." [11] And then again, "In a block universe, as we have seen, past, present, and future do not apply to physical events, and so they neither come into existence nor cease to exist—they just are."[12]

And Grünbaum makes the same point: ". . . our final concern in the consideration of the time problem is the physical status, *if any,* of 'becoming.' Our earlier characterization of the difference between the two directions of time does not, as such, affirm the existence of a *transient,* threefold division of events into those that have already 'spent their existence,' as it were, those which actually exist, and those which are yet to 'come *into* being.' And the relativistic picture of the world makes no allowance for such a division. It conceives of events not as 'coming into existence' but as simply being and thus allowing us to 'come across' them and produce 'the formality of their taking place' by our 'entering' into their absolute future. This view, which some writers *mistakenly* believe to depend on determinism, as we shall see, has been expressed by H. Weyl in the following partly metaphorical way: 'the objective world simply *is,* it does not *happen*. Only to the gaze of my consciousness crawling upward along the life [world-] line of my body does a section of this world come to life as a fleeting image?'[13]"[14]

Perhaps it is because relativity theory employs the static or spatialized time of earlier and later—a view which excludes the process of Becoming—that Minkowski's enthusiastic hope and expectation about one particular consequence of the special theory has not come to pass. He voiced what he felt would be the ordinary view of all future generations: "The views of space and time which I wish to lay before you have sprung from the soil of experimental physics, and therein lies their strength. They are radical. Henceforth space by itself, and time by itself, are

[11] G.J. Whitrow, (1975), *The Nature of Time;* p. 134.
[12] Ibid., p. 143.
[13] H. Weyl, (1949), *Philosophy of Mathematics and Natural Science.*
[14] Adolf Grünbaum, (1964), "Carnap's View on the Foundations of Geometry," pp. 658—659.

doomed to fade away into mere shadows, and only a kind of union of the two will preserve an independent reality.''[15]

Minkowski's mistake lay in his failure to note that the notion of time encompassed a complex group of ideas, and could not simply be tacked on to the idea of space or fused with it. Indeed Einstein stated the issue more problematically when he pointed out: ''For this theory [the electrodynamics of Faraday and Maxwell] . . . showed that there are electromagnetic phenomena which by their very nature are detached from every ponderable matter—namely the waves in empty space which consist of electro-magnetic 'fields' . . .. Since then there exist two types of conceptual elements, on the one hand, material points with forces at a distance between them, and on the other hand, the continuous field. It presents an intermediate state in physics without a uniform basis for the entirety, which—although unsatisfactory—it is far from having superseded.''[16] In effect, not only is the nature of time problematic, so too is the nature of space and of things and fields. Any idea of time and space simply fusing into happy conceptual union is inevitably to be disappointed when the two partners to the union are so unclear.

Time, Flux, and Becoming

There would thus appear to be several concepts of time and several concepts of space used by thinkers, writers, poets, scientists, for their various purposes and from their various points of view as they approach the world in which we live. It is this multiplicity of conceptualizations that is perhaps at the heart of the problem. Indeed, perhaps our whole relationship to reality is, as Einstein suggests, in an unsatisfactory intermediate state. Or is it so? Might it not equally be that a multiplicity of conceptual frameworks is essential to any adequate mental construction of phenomena? Let us pursue for a moment this latter possibility.

The raw experience of space and time is that of an undifferentiated space–time manifold. We live in a mentally constructed world of action and of change, and as long as we do not try to formulate what it is like—to speak about it, as St. Augustine complains—we are not in too much trouble. It is when we do try to speak about it, whether poetically, or in story development in the novel, or in precise mathematical language in physics, or in the analytical or propositional language of philosophy, that the trouble starts.

[15] From a translation of an address delivered at the 80th Assembly of German Natural Scientists and Physicians, at Cologne, September 21, 1908. In *The Principle of Relativity,* a collection of papers by Einstein and others.

[16] Albert Einstein, (1949), ''Autobiographical Notes,'' in Paul Schilpp (Ed.), *Albert Einstein,* pp. 25 and 27.

There is a choice among many formulations: the undifferentiated and uniform space; the world of things or atoms moving in undefined space; the world of shifting fields of force like moving clouds; the clock time of earlier and later, which does not seem to flow; the flux time in which future, present, and past do seem to flow from one to the other.

At first sight, Being seems closely associated with the atomic clock-time world and Becoming with the force-field flux-time world. And yet such a conclusion seems too easy, premature, too tidy. Conceptually it is as easy to interlace Becoming with Being in the atomic earlier-later world as it is to interlace Being with Becoming in the world of fields and flux.

There is something more to be teased out. There is another distinction which runs through the material I have selected, and which cries out for attention. It is the association of the flux time of future, present, past, with life and with psychological processes and with the "subjective" world; and the association of the clocktime of earlier and later with the physical or "objective" world.

This pair of associations of these two types of time with the "subjective" and "objective" worlds is in fact a matter of current controversy among philosophers of science. Grünbaum, for example, in a dispute with Reichenbach, quotes Bergmann to the effect that "'Now' is the temporal mode of experiencing 'ago'", and goes on to argue that "Bergmann's demonstration here that an indeterminist universe fails to define an objective (non-psychological) transient now can be extended in the following sense to justify his contention that the concept 'now' involves features peculiar to consciousness: the 'flux of time' or transiency of the 'now' has a meaning only in the context of the egocentric perspectives of *sentient* organisms and does *not also* have relevance to the relations between purely inanimate individual recording instruments and the environmental physical events they register, as Reichenbach claims. For what can be said of every state of the universe can also be said, *mutatis mutandis,* of every state of a given inanimate recorder. Moreover, the dependence of the meaning of 'now' on the presence of properties peculiar to consciousness emerges from William James's and Hans Driesch's correct observations that a simple isomorphism between a succession of *brain traces* and a succession of states of awareness *does not* explain the temporal features of such psychological phenomena as melody awareness. For the hypotheses of isomorphism renders only the succession of states of awareness but not the *instantaneous awareness of succession.*[17] But

[17] Cf. W. James, (1890), *The Principles of Psychology,* pp. 628—629, and H. Driesch, (1933), *Philosophische Gegenwartsfragen,* pp. 96—103.

the *latter* awareness is an essential ingredient of the meaning of 'now': The flux of time consists in the *instantaneous awarenesses* of *both* the temporal order *and* the *diversity* of the membership of the set of remembered (recorded) or forgotten events, awarenesses in each of which the instant of its own occurrence constitutes a *distinguished element*.''[18]

I shall have cause to return in greater detail to Reichenbach's arguments in his *The Direction of Time,* and the views of others as well. But perhaps we have proceeded far enough for the moment to establish the true complexity of our endeavor. Not only are we dealing with different concepts of space and time, of Being and Becoming, of points and fields, of past, present, and future, of earlier and later, of flux and of clocks, in the conceptualizations of physics and the objective world, but we shall also have to pay due regard to the possibility of there being different concepts of time for the mental world as against the physical world, deriving from the formulation of our subjective experience as against our objective experience, our experience of the inner psychic world as against the outer material world. I hope to unravel some of these complex interconnections.

Chronos and Kairos

There is, finally, one other perspective on time, which may complete this introduction. It is a view embedded in the two different Greek terms used in referring to time: *Chronos* and *Kairos*.[19] In brief, the distinction between these two terms is that of chronological, seriatim time of succession, measurable by clocks or chronometers—*chronos;* and that of seasonal time, the time of episodes with a beginning, a middle,and an end, the human and living time of intentions and goals—*kairos*.[20]

Professor Kermode makes much of the distinction between these two different meanings in his volume of essays, *The Sense of an Ending*.[21] He traces the impact of the dominance of each conception of time upon the literature of a given period in a society. When the chronological serial conception—*chronos*—prevails, the outlook of the society tends toward the more scientific and logical sense of a world which moves on from one stage to another, and that outlook colors the literature both in content and in the way chronology is expressed in plays and novels. When by

[18] Alfred Grünbaum, op. cit., pp. 662—663.

[19] The terms *ora* and *aion* are also used, but in their common meanings of divisions of the day and of eternity, rather than to refer to time in a general sense as is the case with *chronos* and *kairos*.

[20] Aristotle used this distinction. *Chronos* is dating time; but *kairos* is the time which gives value. ''What happens at the right time [*kairos*—season] is good.'' *De Categoriae* Vol. 1, pp. 107a 8 and 119a 26—37

[21] Frank Kermode, (1967, *The Sense of an Ending*.

contrast the more human cyclical sense of time—*kairos*—prevails, the literature tends to be more highly charged with religious and godly ends, with apocalypse, catastrophe, revelation, redemption, and new beginnings.

Kermode was strengthened in this view by a controversy which had raged for some forty years about the significance of the use of *chronos* and *kairos* in the Bible. The controversy centered on the views principally of Marsh,[22] Robinson,[23] and Cullmann,[24] based on lexical analysis of the New Testament, that two very different ideas of time are expressed by these two terms, and that the difference is a matter of considerable theological import. James Barr,[25] among others, rejects this argument, and attempts to prove that *chronos* and *kairos* are much more interchangeable in biblical usage.

Without attempting to assess the merits of one or the other side of the lexical argument, and without entering into the debate over the theological implications, we can note a most interesting and significant pair of meanings, one of which attaches to *chronos* and the other to *kairos,* which are the general meanings in Greek whatever the Greek biblical usage. In the strict dictionary sense, *chronos* has the simple meaning of time in the sense of a length of time or interval: it is the time that appears in time "measurers" or chronometers; the time that can be numbered on a clock by making it a discontinuous succession of points on a line. *Kairos* is the time not of measurement but of human activity, of opportunity: it is the time which appears in *"Kairos pros anthropon braxu metron exei"*—"Time and tide wait for no man"—and in the name of the Greek deity, *Kairos,* the youngest son of Zeus, and the God of Opportunity. *Kairos* relates to its close cousin *kainos* which signifies new, fresh, newly invented or novel; *kainos* in turn connects with *kinein,* to move, from which comes *kinesis*. In short, the *kairos* family of terms is concerned with the time of movement, with change, with the emergence of the new and with active innovation.

Marsh states the distinction as between "chronological time" and "realistic time." The meaning of realistic time is "times known and distinguished not so much by their place in some temporal sequence as by their content . . . the time of opportunity and fulfilment." [26] This theme is elaborated by Robinson who describes *kairos* as "time considered in relation to personal action, in reference to ends to be achieved in it. *Chronos* is time abstracted from such a relation, time, as it were

[22] J. Marsh, (1952), *The Fulness of Time*.
[23] J.A.T. Robinson, (1950), *In the End, God . . . .*
[24] Oscar Cullmann, (1964), *Christ and Time*.
[25] James Barr, (1969), *Biblical Words for Time*.
[26] J. Marsh, op. cit.

that ticks on objectively and impersonally, whether anything is happening or not; it is time measured by the chronometer not by purpose, momentary rather than momentous.''[27]

The importance of the distinction here expressed lies in its similarity to the distinction we have already seen between the psychological time of past, present, future, in which human experience is located, as against the objective time of earlier and later of the physicist. As I shall try to show, a different perspective appears when we define our psychological and social phenomena exclusively as events, as episodes, and become aware of their form in time, in the time of intention and duration or of *kairos,* keeping in mind always, of course, that episodes occur in space—time. It is also of interest to note that whereas *chronos* has come down via Latin into all the Roman-based languages, *kairos* somehow became stuck and remained in classical Greek only. This linguistic hold-up reflects the greater ease which we feel with emotionally unencumbered chronology as compared with the more anxiety-filled experience of the time which brings human intentions and purposes into sharp focus, with their consequent oscillations between success and failure, catastrophe and renewal, and between life and death.

[27] J.A.T. Robinson, op. cit.

CHAPTER TWO

Time and the Time Arrow

The subject of time is full of puzzles and paradoxes and of definitions of many different types of time, many of them mutually contradictory. In this chapter I shall describe a number of these puzzles and paradoxes, to give a picture of present preoccupations with the nature of time. Then I shall critically examine one paradox and a puzzle.

The paradox is the one set by McTaggart: it purports to show that time does not exist. He separates out time seen as moving from future to present to past, from time seen as the fixed relationship between earlier events and later events. He then proceeds to show how, according to his analysis, neither type of time can exist. I shall, however, use McTaggart's formulation to show that something akin to both types exist and need to be taken into account.

The puzzle I shall consider is that of whether or not time is directional, whether it flows, whether there is an arrow of time: if not, how is the sense of flow or passing of time to be explained; and if so, is the arrow of time asymmetrical (unidirectional) or not? The reason for selecting these two problems is that considering them will help to sort through the dense and tangled conceptual undergrowth of thinking about time, and make it possible to lay the foundations for a systematic analysis of the nature of time and the meaning of direction.

Some Puzzles and Paradoxes

The following puzzles and paradoxes will serve to illustrate the many difficult obstacles which lie in the way of an adequate analysis of time. I shall record them at this stage as a stimulus to thinking, and then leave them in abeyance until we can turn to them in later chapters and use them as a means of testing the validity and usefulness of our chosen formulations.

Zeno's four great paradoxes—Achilles and the tortoise, the arrow, the stadium, and the dichotomy [1]—were created to bolster the Parmenidean view and to demonstrate the impossibility of dealing conceptually with time and movement. These paradoxes have been resolved in many different ways by many different writers—Russell, Quine, and Grünbaum among others. But their beauty is that they raise such fundamental questions that they still remain useful after two thousand years, as testing grounds for ideas about time; and so we shall employ them.

Then there is the colorful preoccupation of philosophers with naval battles, started by Aristotle's felicitous choice of this theme in his analysis of fatalism using the ahistorical prediction that "a sea battle will or will not be fought tomorrow.[2] Again, like Zeno's paradoxes, the implications of this proposition have not proven difficult to deal with by means of modern logic, but it serves as a useful medium for the critical assessment of ideas about time, future, and fate.

Another set of questions concerns cause and effect and determinism. Is the nature of time and things such that, as Laplace is believed to have suggested, if one knew everything at any given moment it would be possible to predict all subsequent events?[3]

On a more limited basis, as Hume put the problem, how is it possible to know that A causes B? Are frequency and recency of association sufficient criteria? Hume left the question unsettled in his own mind, and

[1] For example, Achilles can never overtake the tortoise because each time he gets to where the tortoise was, the tortoise has moved on—or the arrow can never reach the target because at any instant of time it cannot be moving.

[2] *De Interpretatione*, op. cit., pp. 18b and 19a. This theme has led to what must surely be the most colorful series of titles in the literature of philosophy; to name but a few: D. Williams, (1951, "The Sea Fight Tomorrow?"; R.J. Butler, (1955), "Aristotle's Sea-Fight and Three-Valued Logic"; C.K. Grant, (1957), "Certainty, Necessity, and Aristotle's Sea Battle"; J.K. Farlow, (1959), "Sea Fights Without Tears"; A.G.N. Flew, (1959), "Hobbes and the Sea-Fight"; C. Strang, (1960, "Aristotle and the Sea-Battle"; J. Hintikka, (1964), "The Once and Future Sea-Fight."

[3] Laplace in fact did not make any such simple suggestion about the possibility of the existence of such an absolute intellect. The famous passage is: "An intellect which at a given instant knows all the forces with which nature is animated, and the respective situations of the beings which compose it—supposing the intellect were vast enough to subject these data to analysis—would embrace in the same formula the motions of the largest bodies in the Universe and those of the slightest atoms: nothing would be uncertain for it, and the future, like the past, would be present to its eyes."

But most writers fail to take note of the sentence which completes the passage, in which Laplace states his viewpoint about the unending task of increasing real human understanding: "All the efforts of the human spirit in the search for the truth tend to enable it continuously to get closer to the understanding of which we have just conceived, *but from which it must always remain infinitely far* [my italics]. This tendency which characterizes the human species is that which renders it superior to the animals, and its progress in this regard distinguishes nations and historical periods, and constitutes their true glory." Laplace (1820), "Introduction à la Théorie Analytique des Probabilités," in *Oevres Complètes,* Vol. 7, p. VII.

Bergson also has shown the fallacy in the narrow deterministic viewpoint, pointing out that our awareness-in-flux of the immediate ongoing situation can be formulated only retrospectively, and by the time it is formulated for predictive purposes, life and the world have moved on. Henri Bergson, (1910), *Time and Free Will.*

subsequent events have made the issue a matter of intense uncertainty; especially, for example, when questions are raised on such politically and socially sensitive topics as the possible inheritance of ability or intelligence.

Relativity theory has thrown further confusion into the cause and effect arena, by raising entirely new questions about the meaning of simultaneity and succession: for example, how is it possible to know whether an event A is anterior to, and possibly the cause of, an event B? By showing that both measuring sticks and clocks change with changes in velocity, we make the measurement of length as problematic as the measurement of time, and the very notion that there might be simple atomistic causes and effects is called into question.

In addition, important new paradoxes have arisen, such as the paradox of the twins. What happens if one of a pair of twins remains on earth while the other takes a trip into outer space at a speed near the speed of light and returns several years later (according to the terrestrial clock and calendar)? Will the space traveler on his return be older or younger than his twin? This paradox bids fair to replace Zeno's paradoxes as a central focus for philosophical debate.[4]

McTaggart and the Nonexistence of Time

The foregoing puzzles about time, while having an inherent interest in their own right, are presented together to give a sample of the uncertainties in the arena of thought about time. I shall consider each one later, at a point where it is germane to the development of my argument. I now propose to turn to one paradox which may prove of great practical value in teasing out some of the main issues from the welter of questions raised by the foregoing paradoxes, puzzles, and definitions. It is McTaggart's Paradox, formulated in 1908,[5] which purports to demonstrate the unreality of the existence of time. This paradox has stimulated much fundamental debate about the nature of time. As Richard Gale has said, "McTaggart's argument is fallacious, but it is fallacious in such a fundamental way that an adequate answer to it must supply a rather extensive analysis of the concept of time, along with a host of neighbouring concepts that are themselves of philosophical interest, such as change, substance, event, proposition, truth, and others." [6]

It will certainly not be my endeavor to tackle these profound and difficult philosophical questions. McTaggart's paradox has in fact been

[4] See, for example, Henri Bergson, (1965), *Duration and Simultaneity*, pp. 163—172; A.N. Whitehead, (1923), "The Problem of Simultaneity."

[5] J.M.E. McTaggart, (1927), *The Nature of Existence*, Vol. II, Book V, Chapter 33.

[6] Richard M. Gale (Ed.), (1968), *The Philosophy of Time*, p. 65.

answered, and I shall briefly review both the paradox and what seem to me to be the most cogent arguments which have exposed the fallacies in his argument. Nevertheless, there is an analysis of two major aspects of time that is more clearly formulated in McTaggart than anywhere else. It is this double formulation that I wish to retrieve, for one or the other aspect tends to be thrown away by those who undo the paradox. I will argue that the two aspects are separate and both are necessary to an adequate formulation of time, but that McTaggart's actual formulations of them were unsatisfactory. That is to say, whereas McTaggart formulates two types of time and proceeds to throw them both away, I shall do precisely the opposite and shall keep them both. With some modification, they can be used to lay the foundation for the two-dimensional theory of time that I shall pursue.[7]

The starting point of McTaggart's argument is to recognize the existence of two main formulations of the nature of time. The first is the time series in which events are first described as in the future, and then as moving from the future to the present, and then through the present into the past—referred to by McTaggart as the A-theory of time, or the A-series. The second is the time series in which one event is seen as happening later or earlier than another, after it or before it—the B-theory of time, or the B-series.

McTaggart then argues (and one of my main points will be to disprove a key feature of this view) that the time relationships in the A-series are constantly in flux, constantly changing, constantly on the move, the future becoming the present becoming the past. Before Julius Caesar was born the event was in the future, once he had been born the event had moved from the present (the actual event) to the past (the A-series). But that Julius Caesar died before Isaac Newton is an unchanging relation in time (B-series): it was always so and will always be so.

McTaggart's viewpoint is that it is only the A-series that is truly temporal: it contains change. By contrast, the B-series is static, and can take on the temporal quality of change or movement only if it can shown to be related to the A-series. The crucial point in his argument is that if the A-series can be shown not to have existence, the B-series also falls as a temporal phenomenon, and hence time is unreal.

In setting out to prove that the A-series cannot exist, McTaggart argues that every event must have mutually incompatible A-determinations—past, present, or future (it is the fact that they are compatible and

[7] This view is held by Mink, about whom Gale remarks: "The only philosopher who has defended the view that neither the A- nor the B-Series alone is sufficient to account for our conception of time is L.O. Mink." R.M. Gale, *The Philosophy of Time,* p. 506; and Louis Mink, (1960), "Time, McTaggart and Pickwickian Language," pp. 252—263.

not contradictory that is the starting point for my own analysis). At any given moment an event is future at a moment of past time, past at a moment of future time, and present at a moment of present time—all involving the coexistence of past, present, and future, which he regards as a contradictory state of affairs. McTaggart extends his argument by demonstrating the infinite regress which results if the attempt is made to get out of the seeming paradox by the view that each of the three events, future, present, and past, occur successively in different time series. It becomes necessary to introduce a never-ending complex of time series as more and more events are considered. In short, the A-series and A-determination cannot exist, because it contains what McTaggart regards as the contradiction of the coexistence of past, present, and future. Hence the B-series does not exist, and hence time itself does not exist.

The Reality of Objective or B-Series Time

In the very considerable literature dealing with the McTaggart paradox, among the most trenchant critics are those who maintain the reality of time as an objective datum—the reality in effect of the B-theory of time, of a time without need for worrying about the psychological complications of past, present, and future, a time which can be dealt with as a clock-time series. Among those who hold such a view, and with different emphases, are Bertrand Russell, Braithwaite, Gale, and Williams.[8] They encompass a range of ideas, and not all their arguments need be elaborated here.

The force of their common outlook is that the B-series time, or what tends to be called real time, exists, and that it is temporal, regardless of whether or not A-series time exists. Included within this argument is the view that this B-series or real time is asymmetrical, that is to say, unidirectional. From this point of view the A-series can be disregarded, it being considered as "merely" a psychological phenomenon; that is to say, the A-series is simply a matter of the cognition of an observer at a given moment in real time of certain remembered events, combined with certain current perceptions, combined with certain current ideas of what the observer would like to see happen or desires or thinks might happen—a combination of complex experience all in the present. This conception I shall return to later in the chapter.

Russell argued, for example, (ideas which he initially formulated before McTaggart's presentation) that change, time, and motion do not depend on what might be called the A-series experience. All that is

[8] Bertrand Russell (1903), *The Principles of Mathematics:* R. Braithwaite, (1928), "Time and Change"; R. Gale, (1967), *The Language of Time;* Donald Williams, (1951), "The Myth of Passage."

required is to reject such notions as *states* of motion, *states* of change, or *states* of velocity. That is to say, there is "no transition from place to place, no consecutive moment or consecutive position, no such thing as velocity except in the sense of a real number which is the limit of a certain set of quotients. . . . Motion consists merely in the occupation of different places at different times." [9]

This handling of time in terms of events occurring in space and time is developed fully by Donald Williams in "The Myth of Passage".[10] As he puts it, "the universe consists, without residue, of the spread of events in space–time, and if we thus accept realistically the four-dimensional fabric of juxtaposed actualities, we can dispense with all those dim non-factual categories which have so bedevilled our race . . .. The theory of the manifold is literally true and adequate—true, in that the world contains no less than the manifold; adequate in that is contains no more."

From this point of view Professor Williams demonstrates that the notion that there is a passage of time, or that time itself somehow flows, is an unnecessary conception. But to get rid of the notion of time flowing (McTaggart's A-series) does not eliminate the concept of time. All that is required is to note that events have extension not only in space but in time also—that is to say, that events are always continuously extended in the space–time manifold. It is precisely this continuous extension which constitutes the very essence of events.

For purposes of scientific study, or other reasons, intellectual abstraction can be made from the space–time manifold—length, and masses, and distances, and time intervals—and changes in spatial position with time (motion, velocity, acceleration) can all be established as conceptual or mathematical ideas.[11] But such mental abstractions and logical symbolic manipulations do not destroy or interfere with the actuality of the continuous extension of events in the space–time continuum—raw experience occurs in the space–time manifold only, and never in space or time alone.

It is this continuous extension which establishes the real time of earlier and later. Professor Williams points out that there is nothing more

[9] Bertrand Russell, op. cit., pp. 445—447.
[10] Donald Williams, op. cit.
[11] This notion is trenchantly stated by Cassirer. In a section entitled "Space and time as mathematical ideals," he writes: "The absolute space and the absolute time of mechanics involve the problem of existence just as little as does the pure number or arithmetic or the pure straight line of geometry . . .. Galileo emphasised most sharply that the general theory of motion signified not a branch of *applied* but of *pure* mathematics. The phoronomical concepts of uniform and of uniformly accelerated motion contain nothing originally of the sensuous properties of material bodies, but merely define a certain relation between the spatial and temporal magnitudes that are generated and related to each other according to an ideal principle of construction." Ernst Cassirer, (1923), *Substance and Function and Einstein's Theory of Relativity,* p. 181.

unusual about the fact that the time-extension in the space–time manifold appears to be directional giving earlier and later, than about the fact that spatial extension is directional also. "In the line of individual things or events, a, b, c . . . z, whether in space or in time, the 'sense' from a to z is *ipso facto* other than the 'sense' from z to a. Only because there is a difference between the ordered couple a–z and the couple z–a can we define the difference between a symmetrical and an asymmetrical relation." [12] I shall show later, however, that this particular conception of directionality from earlier to later is unnecessary.

Karl Popper's conception of objective nonentropic physical processes (events) with an objective arrow of time is similar in orientation to Professor Williams's formulation of "sense" of the extension of both space and time in the space–time manifold. Popper is particularly impatient with any idea that objective, real time of earlier and later does not exist.[13] The reality of objective time does not need to be derived from the existence of the McTaggart A-series. It exists independently in the reality of events in space–time. (See next section).

There are many other arguments which destroy McTaggart's paradox and his argument for the nonexistence of time,[14] and which demonstrate not only that the B-series of time does exist as real or clock time, but also that McTaggart's arguments that internal contradiction inheres in the A-series are false. I do not intend to review all these arguments here. But there is one point that I should like to establish, a point that is raised by an underlying fallacy in McTaggart's description of the A-series, a fallacy which often enough appears.

McTaggart formulates the A-series in terms of an event which exists in the future, then moves into the present, and then becomes past—like an entertainer in the wings of a theater waiting and ready to come on stage, do his act, and then retire to the wings again, his career then being finished, but living on in memory of his one great moment. In fact no such event has ever existed. No event can be said to exist in the future or ever to have existed in the future. We can *never* know what *will* exist, any more than we can know the *ding-an-sich* of the noumenal world. The birth of Julius Caesar could no more be said to exist in the future in 200 B.C. than the birth of goodness knows who in 2200 A.D. can bc said to exist in the future as I write now. To say that it does so exist is to

[12] Op. cit., p. 268

[13] Karl Popper, (1956), "The Arrow of Time," p. 538.

[14] These include C.D. Broad, (1938), *An Examination of McTaggart's Philosophy;* W. Van Orman Quine, "Mr. Strawson on Logical Theory"; J.J.C. Smart, (1963), *Philosophy and Scientific Realism;* L.S. Stebbing, (1936), "Some Ambiguities in Discussions Concerning Time"; Stuart Hampshire, (1965). *Thought and Action.*

return to the mechanistic and absolute determinist world which is so devoid of life that it would not really matter one way or the other whether anyone were born at all in 2200 A.D.

All that we can know about Julius Caesar is that we have records which suggest that he was born in 84 B.C., later than all events which occurred before that date and earlier than all events which have occurred or will occur since that date. And as for what may happen in 2200 A.D., all we can know is a wide range of views stated by individuals, each different from the other, about what is currently thought *might* come to be. These views can be recorded, and the recorded statements can in due course be compared with what is then perceived to be happening. And when that happening has run its course, it is a misnomer to say that it has passed into something called the past. In fact what has occurred is that some records, artefacts, and memories exist as part of the only world we can know—the world of now.

Indeed, even the comparison of a predicted event at the time it happens with the recorded prediction is a grossly watered-down comparison. It is a comparison only of the objectively recordable aspects of a prediction, on the one hand, and of an experienced and recorded event, on the other. It must inevitably leave out of account all the subtle and unformulated (and unformulatable) aspects of the sense and meaning of the prediction at the time it was made, and of the occurring event. Such comparisons are thus at best exceedingly crude—crude, that is, if we are interested in the meanings of human life and experience.

Mink has realized that the true resolution of the McTaggart paradox is to recognize that it is necessary to keep hold of both the A-series and the B-series.[15] He saw clearly that the two series reflect two different ways of looking at the world: what he calls the *discursive* aspect, which throws up our tendency to fix on points, to perceive things in succession, as earlier and later, and hence to construct a B-series; and what he calls the *transient* aspect, in which we have a sense of a moving series of A-series (which he constructs as a series of vertical past-present-future lines), this series of transient and changing A-series being identifiable as earlier and later than each other, and hence giving the sense of succession that goes with the B-series.[16]

[15] Louis Mink, op. cit.

[16] Mink, op. cit., pp. 262 and 263. His view is worth quoting at some length: "The reason why it [McTaggart's critique of time] rises again and again from its own ashes is this: we find it necessary to think about the world in two different ways which we can neither simultaneously express nor separately perfect. Hence it has been both easy and fatal for critics to accept McTaggart's assumption that the concept of time can be discursively exhausted . . ..

There is one way of thinking about the world—and this way is natural and inevitable—in which the world-process is regarded as a sequential manifold of events, strung out along a

The Arrow of Time: Asymmetry and Unidirectionality

If, then, like McTaggart, we can believe that there are things and events which as of *now* actually have existence in the future, which then become present, and which then continue somehow to exist in a disappearing past, we should not wonder that we run into paradoxes—which, like all paradoxes, are not part of nature but are expressions of the shortcomings of our inadequate and confused reflections upon nature, of our construction of knowledge, and of our concepts and language. I propose to develop this theme, but before doing so let me turn to the second of the two puzzles I have chosen; namely, does time move like an arrow in one direction, or in two or more directions, or perhaps in no direction at all?

One type of argument about unidirectionality that has gained currency is that formulated by Reichenbach [17] in terms of entropy. Because, he argued, we live in a world of positive entropy, time must be unidirectional. He drew attention, however, to the possible flaw in this view, which had been suggested by Boltzmann and by Schrödinger (see below), namely that there was no way of proving that entropy was a constant

dimension of occurrence and jointly constituting the world of fact. It is not that the intellect (or syntax!) spatializes time.' Rather we construct a model, whose relations are B-relations, as the formal image of our sense of the particular connexivities of experience. Memory and the experience of the specious present are no doubt conditions of the construction of this model, but it does not represent either or both of them. The chronology of dates is a convenient device which permits us to use this B-model without having it totally in mind at every moment. So it is that we could in principle dispense with chronology if we never thought of an event except as earlier or later than other events . . . .

But there is another way of thinking about the world—and it is equally natural and inevitable—in which the past seems not so much a corpse as a wraith, not so much fixed, as by *rigor mortis,* as irretrievably lost, like old yearnings. What has been is not merely preserved in the amber of settled fact but also cancelled by time's ingratitude. That only the present is real is not just a theory of interest only to philosophers doing philosophy but an expression of this way of thinking. If memory is the matrix of the manifold of events, it is also the heir of *le temps perdu,* because only that can be recollected which can no longer be possessed. Living in the vanishing present, we recognize that past and future are not just other presents, and that even if the past were to be repeated or the future clairvoyantly experienced it would be as a new and different present. The imagination can range at will over the manifold of earlier and later, but it cannot treat in the same way the metamorphosis of past, present and future.

The first way of thinking fixes our attention on the *discursive* aspect of time, the second on its *transient* aspect. Because of the first, we can diagram time's narrow arrow; because of the second, we can recognize time's ticking nick. The nerve of McTaggart's argument is that if we adopt the first way, we can never say what we mean about time, whereas if we adopt the second way we can never mean what we say. And we can adopt neither without the guilty recognition of the claims of the other. The critics of McTaggart have succeeded mainly in providing a continuing demonstration that in this he was entirely correct.

But the 'unreality of time' does not follow from such an admission, *unless* it is assumed that time must share *all* the characteristics of discourse. Nor is it necessary to urge, with Bergson, that 'real' time has properly no discursive characteristics at all. Because time does have discursive characteristics, we do in part mean what we say and say what we mean. But because it has also a transient aspect, I think it necessary to admit that our discourse *about* time is essentially Pickwickian, in the special sense that it is the use of language rather than the meaning of its terms by which we communicate the sense of temporal transience."

[17] Hans Reichenbach, (1956), *The Direction of Time*.

condition. The world might just be in a phase of positive entropy, which could just as well be followed by an upswing phase of negative entropy. In this latter case, it could be argued that the time arrow too would be reversed.

In order to get around this problem, Reichenbach tried, by an analysis in terms of branch entropy systems,[18] to establish that time would remain asymmetrical regardless of the general condition of entropy. Popper and Grünbaum have argued, however, that this analysis starts from wrong premises and becomes unnecessarily complex, and have put forward a stronger and more coherent case.

Popper [19] bases his case upon what he criticizes as the subjectivist's view of entropy of Boltzmann and Schrödinger from which they derived the possibility of time's being able to run in two directions, depending on whether the general entropic state of the world was in a positive or a negative phase.

Popper's view, and strength of feeling about it, comes through in his criticism of Boltzmann's reply to Zermelo,[20] which he expressed as follows:

"I think that Boltzmann's idea is staggering in its boldness and beauty. But I also think that it is quite untenable, at least for a realist. It brands unidirectional change as an illusion. This makes the catastrophe of Hiroshima an illusion. Thus it makes our world an illusion, and with it *all our attempts to find out more about our world*. It is therefore self-defeating (like every idealism). Boltzmann's idealistic *ad hoc* hypothesis clashes with his own realistic and almost passionately maintained anti-idealistic philosophy, and with his passionate wish to know.

"But Boltzmann's *ad hoc* hypothesis also destroys, to a considerable extent, the physical theory which it was intended to save. For his great and bold attempt to derive the law of entropy increase ($ds/dt \geqslant 0$) from mechanical and statistical assumptions—his H-theorem—fails completely. It fails for his objective time (that is, his directionless time) since for it entropy decreases as often as it increases. And it fails for his subjective time (time with an arrow) since here only a definition or an illusion makes the entropy increase, and no kinetic, no dynamic, no statistical or mechanical proof could (or could be required to) establish this fact. Thus is destroyed the physical theory—the kinetic theory of

[18] H. Reichenbach, ibid. He attempted to demonstrate that if there were alternations between epochs of positive and negative entropy in the universe, then concomitant alternations would occur in the direction of entropy increase in the branch systems, leaving the direction of time unchanged.
[19] Karl Popper, (1974), *The Philosophy of Karl Popper;* and (1976), *Unended Quest*.
[20] L. Boltzmann, (1964), *Lectures on Gas Theory*, pp. 446ff.

entropy—which Boltzmann tried to defend against Zermelo. The sacrifice of his realistic philosophy for the sake of his H-theorem was in vain." [21]

Popper thus seeks to get away entirely from the notion of a moving arrow of time, while retaining the concept of unidirectionality in the sense of earlier and later. That there is a real time of earlier and later, yes, but as far as any idea of a *flow* from earlier to later is concerned, no such conception is necessary. This view is like Williams's notion of there being a sense in the space–time manifold, but that movement in a direction is not a necessary idea. For if there is flow in a direction, then at what rate, and in what direction? That these questions cannot be given meaning, not at least in connection with the so-called real time of earlier and later, is a view which would seem self-evident. The question of the directionality of past, present, and future, however, I leave for later consideration.

Grünbaum [22] makes the same point as Popper, formulating the notion of earlier and later in terms of the anisotropy of time rather than any direction of flow of time—of objective or real time, that is. He does, however, allow for a particular meaning to direction in the psychological experience of time, the direction into the felt future. I shall take further this latter notion in terms of the direction of intentionality rather than the direction of time.

Finally, a telling coverage and analysis of this problem of the symmetry or asymmetry of time is to be found in *A Treatise on Time and Space* by J.R. Lucas.[23] Lucas avoids all the complications of the entropy theory by going directly to an analysis of the direction of events in the time dimension of the space–time manifold in terms of communication between human beings. He draws attention to three main conditions. The first condition is the causal means of communication. For two people to communicate they must be sending and receiving responses in the setting of a uniform direction of time; otherwise, for example, one person's answer would precede the other's question, and their interaction would get increasingly out of gear as it progressed.

Second, behavior and communications are intentional. If time were symmetrically bidirectional, then if two people were functioning in a direction opposite to the normal direction of time, the intentions of the one would appear as causal antecedents and not the intended consequences of his communication. In effect, writing would appear to be erasing. And third, the logic of each one's communication would seem to be reversed. In the first place, the other's sentences would seem to be backwards. And even if they could be unraveled, no argument or discussion could take

[21] Karl Popper, *Unended Quest,* pp. 161—162.
[22] A. Grünbaum, op. cit.
[23] J. R. Lucas, (1973), *A Treatise on Time and Space.*

place. The progression of thought in each case could be in the wrong direction.

Lucas thus pins himself to a unidirectional conception of time. He states, "Directionless time is not time at all. Everybody believes that time has a direction, but many allow themselves to be confused about it. . . . If we consider more carefully what is involved in two beings communicating with each other, we can see that it is a necessary condition of their being able to communicate that they should experience time going in the same direction. A uniform direction of time is therefore an essential condition of intersubjective experience."[24]

Finally, at the end of his chapter on the asymmetrical nature of the directionality of time (Chapter 8), Lucas, in contrast to Williams, makes a comparison between space and time, arguing that "space is necessary to give us room to be different in, to give things opportunity to change—and to change back again. These are symmetrical relations, and so space is undirected, and lacks this quintessential characteristic of time." What is of importance to us here, however, is their agreement on the asymmetrical sense and direction of time, despite their difference on whether there is any inherent sense in space. I shall show later that this agreement is founded more upon our experience of direction of events than of time.

Time and Direction: Physical or Psychological Facts

The material which I have selected gives some considerable support to the idea of B-series time as a reflection of the external world and to the idea that events in B-series time have a relationship of earlier and later, and that these relationships are asymmetrical or anisotropic. There is much less agreement about the existence of A-series time in the external world and about time having direction in the sense of there being a passage or flow or arrow of time moving along a particular path or trajectory.

The general tenor of arguments in favor of A-series time and the concept of temporal direction is that they are both phenomena from the psychological world of experience, of memory, perception, and expectation, and of direction of intention in the sense of goal-directed behavior. Sometimes this viewpoint is rather more implicit than explicit, as in Lucas. Whether this psychological A-series time is any less real than the "real" time, as Popper calls it, of the B-series is a question which we shall have to consider in detail.

For our purposes, I shall take it that McTaggart in his paradox set out two major modes of experience of time, but that he made two serious errors. First, he was wrong in arguing that B-series time was not in itself

[24] Ibid.

temporal; he used the false criterion of temporal flow for judging temporality. Second, he was wrong in arguing that A-series time contained internal contradictions; his conception that events flowed from future to present to the past was mistaken.

I have found it useful to take McTaggart's analysis and stand it on its head. Instead of getting rid of both the A-series and the B-series, we can retain *both* these conceptions as Mink did and take note of their emergent properties. The B-series can then be seen to define the time of the physicist's space–time continuum; it is anisotropic and asymmetrical, and it contains earlier, simultaneous, and later, as statements of sense relationships in the space–time field. By contrast, the A-series can then be seen to be the time of the psychologist's introspective subject and arises from that subject's sense of past, present, and future (but divested of any notion of anything moving from future to present to past).

These two ideas of time encompass all of the times described in this chapter. The physical concept of time encompasses the ideas connected with objective time, with real time, and with external time: it includes the concepts of later, simultaneous, and earlier. The psychological concept of time can absorb the ideas connected with physiological time, subjective time, and intuitive time: it includes past, present, and future times.

The various ideas connected with the flow of time, or arrow of time, can be dropped with respect to the B-series in favor of the single concept of time as *chronos* with relationships of sense between events in a field, of the extension of events in a field. That is to say, given two events recorded as having occurred, statements can be made as to which occurred before the other, without having to import any conception of time having flowed from one to the other like an arrow with a flight path aimed in a particular direction, flying from the earlier toward the later. By contrast, as far as the A-series is concerned the problem takes on a different connotation. To the extent that the concept of direction remains, it does so transformed into the concept of direction of intention and striving with time as *kairos* with which the A-series is connected; directionality lies in the flow of behavior and of the psychic experience of events rather than the flow of time.

Finally, we shall leave for the moment a number of ideas such as calendars and clocks on the one hand, and the different kinds of present which have been suggested by various authors, on the other. How these recording and measuring instruments can be used with respect not only to the physical idea of time but also to the psychological idea of time will prove worthy of being treated as a subject in its own right. And so too will the concept of the present and its nature. I shall pick up some of these themes in the next chapter.

CHAPTER THREE

Time as Continuity and Change

The literature on time suggests innumerable different kinds of time: subjective time, real time, objective time, linear time, cyclical time, proper time, absolute time, relative time, external time, internal time, biological time, psychological time, clock time, intuitive time, and more besides.[1] Our analysis thus far, however, suggests that there might be two major kinds of time; or if not different kinds of time, then at least two main ideas about time, or concepts of time.

In this chapter I shall first call attention to the fact that time as a thing-in-itself, as a particular, is a general idea and is not an agent. It cannot do anything or cause anything to happen. I shall then consider further the question of how many types of time there might be: the question of whether we really need more than one concept of time to make sense or order out of the range of experience which gives rise to the idea of time.

The conclusion I shall reach is that only one concept of time is needed. This concept can be applied regardless of whether we are treating of external objective ponderable experience or of internal subjective imponderable experience, thoughts, imagination or dreams, earlier or later, past, present, or future. It is definable in terms of the conceptual organization of the experience of change, of process, of motion, of transformation, as well as of object constancy and permanence experienced as continuity of existence. This definition of the concept of time is extended by reference to the operational definition of length of time, and is compared with the concept of space.

This univocal concept of time is not, however, the same as either of the two "types of time" we thought we had identified—namely, the

[1] The most comprehensive survey of the difficulties associated with clarifying the nature of time, is to be found in the collection of essays, with an outstanding commentary by the editor, J.T. Fraser (Ed.), (1966), *The Voices of Time*.

physical and the psychological. These two concepts turn out to be not different types of time but rather ways in which human behavior is organized in relation to time. They can best be seen as two axes or dimensions of time, analogous to the coordinates of space.

The Use of the Word 'Time'

Just as is the case with all language, we can use the word 'time' as a universal or category and as a particular. Thus, just as we may refer to chair, or man, or space, or triangle, or idea, as universals, as categories, we may refer also to time as a universal or category. That sense is the one most commonly considered in philosophy. But it is also the case that just as we may refer to 'that particular chair,' or 'that particular man,' or 'that particular triangle,' or 'that particular idea,' we can just as readily refer to 'that particular time.'

It is in its use as a particular that some of the confusion about the concept of time can arise. The word 'time,' like all language (except syntactical connectives), can be formed into noun, verb, adjective, or adverb. As a noun, we can speak of "the time it happened," or "the time it took," or "the time when Imperial Rome was ascendant." As a verb, we can speak of "the man timing a race or a game," or "the scientist timing a process." As an adjective, we can speak of "a timed event," or "a time machine," or "a temporal quality," or "a well-timed blow." As an adverb, "timely' has fallen into disuse (or been taken over as an adjective as in "a timely event"), but common enough archaic usage is illustrated by Macduff who was "from his mother's womb untimely ripp'd."

But because we can make full use of time as a particular we tend to confuse it with the most common types of particular, namely words which refer to concrete things—whether physical or psychological or social things—which can do something, which can cause something to happen, which can act as agents. Examples of such concrete agents are legion: a shooting star, an explosion, a river, a stone, a hypothesis, a political slogan, an attitude, an insult. But time is not such a word. It does not refer to a concrete agent. It does not refer to something which can have consequences in the sense of agency. Time does not do things.

If it does not refer to a concrete agent, to which sort of thing does time, used as a particular noun, refer? It refers to a positional idea, a way of ordering things and events by the vastly important human process of positioning them by dating them, putting dates upon them, putting time tags on them. When we say, "The time of the marathon race is three o-clock, Tuesday, November 28, 1978," we date it. When we say, "The time for the race was 3 hours, 47 minutes, 32 seconds," we mean that if it started on time at three o'clock, it finished at 47 minutes and 32 seconds past six o'clock that same afternoon, thus dating the course of the whole event as an event identifiable and specifiable in time, as an episode which can be dated by the simple means of retrospectively dating its beginning and its end.

The distinction I am making is that between agency particulars, like physical or psychological objects, and particulars of state or position which are not themselves objects but which can be used to help to give a sense of order to our picture of the world. Examples of particulars of state or position, in addition to that of "the time," are "the form of something" or "the position of something."

The reason for my seeking to make clear this distinction between positional nouns such as time, and agency nouns such as stick or person, is to get away from the confusion that arises if and whenever time is used as though it were an agency noun, that is to say, used as though it referred to an object or idea which can act as agent. It is in this incorrect sense or usage that we refer to "time flowing," or "the passage of time," or "the arrow of time," or to "the directionality of time" or "the reversibility of time," as though there were something concrete which could fall like a stone, or flow, or fly, or pass us by, perhaps joining us and going our way or perhaps turning around or reversing like a winged horse and going off in some other direction.[2]

We make the same mistake with other state or positional or ordering nouns. Because we can so easily say that "the position of the car changed" or "the form of the clouds changed," we can equally readily believe that form or position can in and of themselves change and, in so doing, somehow change the car or the clouds. In fact, of course, the particular agents are the car and the driver, or the wind and the vapor, the position and form refer to our mode of organizing our observations of what we judge to be going on in the world.[3]

I shall have cause to return to this subject of time as an ordering noun, and to show that our attempts to treat time as a particular as though it referred to a concrete agent are akin to, and perhaps nothing more than, the common human tendency toward anthropomorphism. But let me leave

[2] Whitehead was very sensitive to this distinction. As he put it on one occasion, "The measurable time of science and of civilized life generally merely exhibits some aspects of the more fundamental fact of the passage of nature." A.N. Whitehead, (1920), cf2The Concept of Nature, p. 54.

[3] "Homogeneous space and homogeneous time are thus neither properties of things nor essential conditions of our faculty of knowing them: they express, in an abstract form, the double work of solidification and of division which we effect on the moving continuity of the real in order to obtain there a fulcrum for our action, in order to fix within it starting points for our operation, in short, to introduce into it real change. They are the diagrammatic design of our eventual action upon matter." H. Bergson, (1911), *Matter and Memory,* p. 280.

this matter for now, and return to our problem of how the idea of time arises, and whether there is but one kind of time or many.

The Experience of Time and Events

As we have already seen, there have been many different descriptions of the nature of the experience which gives rise to the notions of time or space, or, contrariwise, the kinds of experience or organization of experience which arises from time and space as *a priori* categories. The one common element in these descriptions is that neither time nor space *per se,* as such, in their own right, can be directly observed, touched, seen, felt. They have somehow to do with our sense of extension of change, of things located in relation to one another, and changing both in relationship and form, so that somehow things are always either different or—what amounts to the same thing—apparently remaining the same: to say that something remains the same is as much a statement about that thing at at least two different times as the statement that something has changed.

What then are the experiences which might be linked to time? Before looking at this question, let me establish one simple point; namely, that in speaking about time we are not only speaking about an abstract noun but, as in the case of all other words, we are speaking about a verbalized idea, a concept, an abstraction from experience and a putting into words. We need not dwell on this point, for the idea that the word is not the thing, the concept is not the referent or image (as Frege would put it), calls for no argument. It is just that in the case of time—as in the case of space—it is worth recalling that we are dealing with mental abstractions. The import of so doing is that it serves to give some sense to the idea of asking not, ''What then is time?'' but rather, ''What then are the kinds of phenomena which we endeavor to organize within the general construct of time?''

The question to be explored, then, is to what extent such differences as appear in the descriptions of time—such as real time, psychological time, biological time, objective time, subjective time, internal time, external time—refer to different kinds of time, and to what extent they are simply statements of the time characteristics of different real, psychological, objective, subjective, internal or external things or events, or experiences of such things as events.

There are, in the first place, many kinds of common-sense time. One of these is what might be termed notched time, showing how much time has passed—how many days or moons or seasons. Notched time grows readily into a more general calendar time, which notches up ''future'' dates and periods as well as past ones. Calendar time may be perceived as either cyclical—the recurring seasons, for example, or as serial—the time in which things are seen not as recurring but as progressing by cyclical unfolding during which things and people get older.

Which view of time—cyclical as against serial—people take up is related in an important way with attitudes toward death. Cyclical time fits in best with those who seek immortal recurrence through reincarnation of souls.[4] Serial time leads on to facing and accepting death as final whether as final on earth with life beyond or simply as final.

Then there is the more sophisticated objective time which appeared with the discovery of the clock. For here it is possible to believe that one can literally see time moving on. Objective time is time which supposedly exists "out there." It is objective in the sense that there are clocks which can be observed in common with others and from which common readings can be agreed. Experience with objective time, and particularly evidence from experimentation which involves time, gives rise to the idea of real time, "real" taking on the meaning consistent with the particular metaphysical outlook—the conception of reality—of whoever is using the term.

As I described in the last chapter, objective time and real time are associated with motion and change, with the idea of things happening simultaneously, or earlier than or later than other things. Three different types of objective time, however, were put forward, depending on one's ideas about directionality. One view was that time flows in one direction, another that it is bidirectional—flowing first one way and then the opposite way as the direction of entropy changes—and the third view was that it is wrong to think of passage, or flow, or arrow of time, and that earlier and later do not necessarily imply any direction at all.

Another division of time is that into *internal* time and events and *external* time and events. We may, for example, be observing two flashes of lightning, or listening to a concert, these events being experienced as external in the sense that they are occurring outside ourselves, out there, and are available to the observation of other people who are present at the same occasion. Or we may be imagining such events, so that they are experienced as occurring in our own minds, inside ourselves, in here, and not available to the direct observation of others.

Is, then, time different in the two situations? It does not immediately appear so. Of the two flashes of lightning, one is experienced as arriving at the observer earlier or later than the other, or as arriving simultaneously with the other, regardless of whether they occur in the external world or

[4] The Buddha, for example, is seen as returning reborn in cycles of time of a duration so long as to be inconceivable to the human mind—cycles in which the fleeting human second is represented by hundreds of millions of years.

are imagined in the internal mental world. And in the same way, the beginning of a piece of music—regardless of whether it is being played by an orchestra out there or by a mentally imagined orchestra in here—will be experienced as occurring earlier than the end, the end later than the beginning, and when instruments are playing notes in unison they will be experienced as blending simultaneously in harmony.

Moreover, when the lightning is occurring or the music is playing, whether externally or in imagination, it is experienced as present: the present of external observation is the same concept of present as is the present of mental events of the internal world. In the same way, events which have been experienced to have occurred will be rememberable as past, the same concept of past applying in both situations. And so also, the same conception of future will apply, whether on the one hand a person is waiting for a concert to start or for a flash of lightning to occur, or on the other hand has decided to set about imagining a flash of lightning or a particular melody. In short, the same conceptions of past, present, and future, and indeed of memory, perception, and expectation, apply to both external and internal events. And as we shall see, the same problems associated with past, present, and future will arise in connection with both types of event.

By the same token, both the external and the imagined events will be experienced as occurring quickly or slowly, as having high or low velocity, as accelerating or decelerating. Even measurement of rate of occurrence of events will be the same—as far as time is concerned, that is. Whether imagined or external, the events will be conceptualized in terms of their lasting minutes or hours or days. True, there are apparent differences between how long the imagined events ''really'' take as compared with the external ones. But that is not a difference in the fundamental conception of time itself: that is a difference in the nature of imagination as compared with perception. When we imagine events we do *not* change the temporal conceptual frame to a different frame from that used when we are observing external events, any more than we shift our spatial conceptual frame. The striking fact is that we retain the same frame of reference in both situations.

We might then consider whether it makes a difference to the concept of time if the objects and events are *ponderable* or *imponderable*. The answer must necessarily be the same for the external and the imagined, since even if we took an event such as an auto race, the same considerations would apply to the racing cars that are externally observed and ponderable as to those racing cars that are internally imagined and imponderable.

But what, then, about *objective* as against *subjective* events? In the first place, we must make up our minds what we mean by this distinction. Are we to mean the same thing as in the distinction between externally perceived and imagined? Such a distinction does not help, since we would merely be referring to external and imagined events by two different names. I propose to use a distinction which can be derived from an extension of Popper's definition of an objective event as one which has been formulated in a manner allowing it to be communicated and shared with others. I propose to refer to any socially communicated event as objective, regardless of whether it was externally observed or internally imagined in the first place. By the same token I shall refer to the same events as subjective if they have been consciously formulated and communicated to oneself but not to others.

Here again, in the case of objective and subjective events, we shall find that no fundamentally different times emerge as a result of objective formulation. All that happens is that we have statements (either kept as private and subjective or made public and objective) of the temporal qualities of earlier, simultaneous, and later, and of the sense of the temporal content of velocity and acceleration, in addition to the raw unstated experience of events.

So far, then, although we find different circumstances in which events may occur, there appear to be no differences in the nature of the concept of time associated with these various circumstances. There are serial and cyclical events, external and imagined events, ponderable and imponderable events, subjective and objective events, yes, but all occurring with the same type of spatial and temporal reference. There would seem to be but one kind of time and one kind of space, whether we are engaged in imagining or perceiving, or in ruminating or communicating.

Even if we consider the concept of so-called real time, the conclusion remains the same. For we are here faced with the questions of what is meant by "real" time. Is it to be contrasted with "unreal" time? Any such contrast is difficult to make sense of. I would assume that the intended distinction is between "real" time—that which I have carefully tried to tease out as connected with external, ponderable, and objective events—as against, perhaps, "psychological" time—that connected with the categories of internal, imponderable, and subjective events. Our analysis suggests that regardless of whether one's metaphysical view of the world is that external physical events are more "real" than events in the internal psychological world, or that events in both worlds are equally "real," the implications for time are the same.[5] There is no "real" time as against some other kind of time. Time is one single univocal conceptual construction—and applies equally to perceived physical worlds or imagined worlds, to external or internal worlds, to subjective and objective worlds.

Indeed, even if we examine what happens in dreams, where space and time appear to be distorted, we will find that our point of reference consciously and conceptually is still the same time. There is no way out. In dreams it is not space and time *per se,* as conceptual abstractions, which are distorted, which behave in odd and peculiar fashion. It is the manner in which events occur which may on conscious retrospection appear strange: the end of a musical performance may occur before the beginning; familiar processes may seem to run backwards; objects may behave in odd fashion; one thing may simply turn into another, perhaps without seeming to have changed. Such things seem odd only if they are placed in a normal space and time framework. It is not, for example,

[5] Ernst Cassirer demonstrates convincingly the incompleteness of both the physical real and the psychological real taken by themselves: " . . the question as to which of the two forms of space and time, the psychological or the physical, the space and time of immediate experience or of mediate conception and knowledge, expresses the *true* reality has lost fundamentally for us all definite meaning. In the complex that we call our 'world,' that we call the being of our ego and of things, the two enter as equally unavoidable and necessary moments. We can cancel neither of them in favor of the other and exclude it from this complex, but we can refer each to its definite *place* in the whole. If the physicist, whose problem consists in objectification, affirms the superiority of 'objective' space and time over 'subjective' space and time; if the psychologist and the metaphysician, who are directed upon the totality and immediacy of experience draw the opposite conclusion; then the two judgments express only a false 'absolutization' of the norm of knowledge by which each of them determines and measures 'reality.' In which direction this 'absolutization' takes place and whether it is directed on the 'outer' or the 'inner' is a matter of indifference from the standpoint of pure epistemology. For Newton it was certain that the absolute and mathematical time, which by its nature flowed uniformly, was the 'true' time of which all empirically given temporal determination can offer us only a more or less imperfect copy; for Bergson, this 'true' time of Newton is a conceptual fiction and abstraction, a barrier, which intervenes between our apprehension and the original meaning and import of reality. But it is forgotten that what is here called absolute reality, *durée réelle,* is itself no absolute but only signifies a standpoint of consciousness opposed to that of mathematics and physics. In the one case, we seek to gain a unitary and exact measure for all objective process, in the other we are concerned in retaining this process itself in its pure qualitative character, in its concrete fullness and subjective inwardness and 'contentuality.' The two standpoints can be understood in their meaning and necessity; neither suffices to include the actual whole of being in the idealistic sense of 'being for us.' The symbols that the mathematician and physicist take as a basis in their view of the outer and the psychologist in his view of the inner, must both be understood as *symbols.* Until this has come about the true philosophical view, the view of the *whole,* is not reached, but a partial experience is hypostasized into the whole. From the standpoint of mathematical physics, the total content of the immediate qualities, not only the differences of sensation, but those of spatial and temporal consciousness, is threatened with complete annihilation; for the metaphysical psychologist, conversely, all reality is reduced to this immediacy, while every mediate conceptual cognition is given only the value of an arbitrary convention produced for the purposes of our action. But both views prove, in their absoluteness, rather perversions of the full import of being, *i.e,* of the full import of the *forms* of knowledge of the self and the world." Ernst Cassirer, (1923), *Substance and Function,* pp. 454 and 455.

that the temporal concepts of earlier and later undergo change. On the contrary. They remain constant. It is the way in which events occur which is unusual, in the context of time as usual.

But, it might be argued, there are some dreams in which time itself seems to go more slowly or more quickly. Here again, it is not some tangible or directly observable thing called "time" which goes slowly or quickly. It is the dream events which do so. For an event to seem slow or fast it must be held in the context of a normal conceptual time frame. A slow-motion film seems slow not because time is moving slowly but because the people or things in the film seem to be moving unusually slowly. Just as when a runner stops running and begins to walk, we say that it is he who has slowed down, not time.

Continuity, Transformation, and Identity

We are led to the conclusion that there is only one kind of time, regardless of the type of situation in which it "occurs." That is to say, there is need for only one construct, one abstraction, one conceptual category, one class or set, within which to order the wide-ranging experiences to which we attach the sense or feeling of time. There may be many subsets, depending on the situation in which the experience occurs, but they are subsets and not different sets or classes. This conclusion would suggest that the ideas of A-series and B-series, of *chronos* and *kairos,* do not refer to different kinds of time, but rather to different types of event and experience, and to different modes of organizing experience in time. In particular, I shall suggest in the next chapter that the differences have to do with two dimensions of time, in the same sense as three dimensions of space.

If, however, there really is but one kind of time, regardless of how many dimensions it might have, then what is this univocal concept connected with? Reflection on the foregoing material points to a familiar notion; namely, that our idea of time is connected with change, with motion, with transformation. In order to consider time characteristics, it is necessary to consider events, objects in motion or in process of change; or to put the matter more systematically, objects in process of change in state, or change in location, or both. We are in the field of transformations, of metamorphoses.

It is in such circumstances of change that we can refer to an event; it is a change in an object from state 0_1 to state 0_2, which we locate at place S_1 and then S_2, at time T_1 and time T_2. It is the experience of the continual succession that is change which is linked with our sense of time.[6]

This view is the view first expounded by Leibniz, who saw time as a function of observed relationships between objects—relationships of simultaneity and succession.[7] This relational view of time was posited against Newton's notion of an absolute time which would flow on even if the universe were empty.

It is Leibniz's view which has proved most consistent with the development of science and modern philosophy. As Whitrow has observed, Leibniz's conception of time is readily integrated even into relativity theory, something which would be impossible for Newton's reified time.[8] As we shall see, however, there is still a fairly widespread tendency to treat time as though it were a thing, a kind of *zeit-an-sich,* and wherever this happens unnecessary confusion is created.

The other side of this same coin is that apparent permanence, continuity, constancy, are just as much raw data for the concept of time as are any successions of changes. (I say ''apparent'' because there is of course no complete permanence: it is at most an approximation.) The observation that no change has occurred in an object requires statements about the state of the object at at least two different points in time. And even in the case of change, there is change only in something which has been deemed to have continued to exist.

This interconnection between permanence or continuity on the one hand and change or transformation on the other, is the main ingredient of identity: that is to say, of the statement that a thing or person is the same thing or person at two different points in time. Identity must be established much as the provenance of a work of art must be established. A work of art can be said to be identical with that created by the artist so long as an unbroken series of owners can be established from the time it left the hands of the artist, even though the varnish may have yellowed and the paint cracked or faded. Identity means that an object is the demonstrable continuation of an object for which there is a valid and reliable record of existence at an earlier point of time—even though

[6] Interestingly, Aristotle assigned both time and space to his category of continuous qualities: but he failed to consider their possible treatment as discrete and discontinuous quantities, as in measurement, and thus overlooked the quality of succession, of earlier and later. ''Space and time also belong to this class of [continuous] qualities. Time, past, present, and future, forms a continuous whole. Space likewise is a continuous quantity.'' From *De Categoriae,* p. 5a.

[7] G. J. Whitrow has neatly summarized this point: ''The idea that moments of absolute time exist was rejected by Leibniz, who argued instead that events are more fundamental. In his view, moments are merely abstract concepts, being classes or sets of simultaneous events. He defined time not as a thing in itself but simply as the order in which things happen.'' G. J. Whitrow, (1975), *The Nature of Time,* p. 86.

[8] Ibid., p. 96.

change and transformation may have occurred along the way. That is to say, even though the object is not exactly as it was, it is nevertheless that unique object which has changed, and not another one. In current phraseology, it would be called a continuant.

T. S. Eliot vividly described this interweaving of permanence and change, or as he puts it, of stillness and the dance:

> At the still point of the turning world. Neither flesh nor fleshless;
> Neither from nor towards; at the still point, where the dance is,
> But neither arrest nor movement. And do not call it fixity,
> Where past and future are gathered. Neither movement from nor towards,
> Neither ascent nor decline. Except for the point, the still point.
> There would be no dance, and there is only the dance.[9]

Definitions of Time and Space

In the absence of any available ready-made and generally accepted definitions of time and space, I shall adopt a relational definition; that is to say, I shall define them as ordering terms.[10]

Time: formulated conception of the experience of events, of the permanence (identity) and transformation of objects; i.e., of objects continuing to exist at different points (identity, $x = x$) in similar or changed state (whether changed in condition, or location, or both): such points may be termed temporal points, the distance between them termed a length of time, and the general sense of the extension of any or all such points and of the distances between them may be termed the sense of time itself.

Space: in similar vein, we may define space as the formulated conception of the experience of the at-the-same-moment (simultaneous) extension of objects and of their at-the-same-moment (simultaneous) location in relation to one another; e.g., a string extended between its two ends each treated as points, or two coins each treated as a point located in relation to the other: such simultaneously related points may be termed spatial points, the distance between the points termed a spatial length,

[9] T. S. Eliot, (1944), "Burnt Norton," in *Four Quartets*,p. 9.

[10] This usage is in fact the most common usage, regardless of the crude tendency to regard time and space as things. A. N. Whitehead expressed it succinctly: "By saying that space and time are abstractions, I do not mean that they do not express for us real facts about nature. What I mean is that there are no spatial facts or temporal facts apart from physical nature, namely that space and time are merely ways of expressing certain truths about the relations between events." A. N. Whitehead, (1920), *The Concept of Nature,* p. 168.

and the general sense of the extension of any or all such points and of the distances between them may be termed the sense of space itself.

These two definitions refer to two sets which are subsets of the more encompassing category of the space–time manifold. How to define this latter concept I shall leave, however, to the next chapter, in which I consider the more primitive experience of the world all-in-a-piece, that is to say, the qualities of space, time, form, and state, not separated conceptually from one another.

We can test these definitions of time and space by reference to a well known paradox, namely, that whereas it is possible for something to be in the same place at two different times, it is not possible for it to be at the same time in two different places. Many philosophers have gone to great lengths and exercised great ingenuity in trying to resolve this seeming paradox.[11] But the whole paradox vanishes if we recognize the simple fact that an object cannot really be in the same place at two different times; in the same spatial relation to some other objects, perhaps, or even in the identical place (i.e., a continuant of the previous place). But certainly not in the same place, for all sorts of relations with other objects will have changed.

The location of a billiard table has often been used as the example of things remaining in the same place. But in fact if the billiard table remains "still," all we can say is that the very limited relationship between it and (perhaps) the room or building in which it is located, has continued. But this fixed relationship (continuity) will certainly not hold true with respect to people, to passing cars, to the rain which has begun to fall, or to the sun, the moon, or the stars. In speaking of "the same place at two different times" all we have done is to notice and select the fixed relationships, and to overlook and suppress all the relational changes and transformations.[12]

It is possible only in the approximation of rough and ready everyday common-sense language usage for things to be at the same place at different times. This kind of approximation will not do for our purposes. In particular, it would make difficulties in connection with the measurement of time. Does the second hand of a watch with a circular face

[11] See, for example, Taylor's complicated verbal gymnastics in which he argues that the same object can be in two places at the same time because two parts of that object (e.g., two opposite ends or sides) are at different places. See Richard Taylor, (1955), "Spatial anxd Temporal Analogies and the Concept of Identity." Lucas also accepts this difference between space and time and attempts, equally unsuccessfully in my view, to extricate himself from the apparently paradoxical formulation. See J. R. Lucas, (1973), *A Treatise on Time and Space*. Chapter 17.

[12] Whitehead and Russell exposed the fallacy in this paradox many years ago. See, for example, Chapter 7 of A. N. Whitehead, (1938), *Science and the Modern World*.

traverse the same cycle of places each sixty seconds; and so too, the minute hand each hour, and the hour hand each half-day?

All that it is possible to say is that there is a move to the approximate reconstitution of very limited conditions similar to those which obtained at an earlier point in time. These very limited conditions apply only to the approximate relationship between the hands of the watch and the watch face. Thus, for example, if the wearer of the watch had "remained in the same place" but the earth had somehow moved into a strong electromagnetic field, we would hardly then refer to the watch as having remained in the same place; the resulting peculiarities in its behavior would draw attention to the change in place relative to the larger surrounding environment.

It might be argued, of course, that the hands kept cycling around the same place, but that the condition of the place had changed as a result of the impingement of the electromagnetic field. In fact, it really does not matter which view is taken. Relativity theory has shown just how much it is a matter of pragmatic choice whether we say that the location of an object or event is fixed and state changed; or state and location have changed; or location has changed and state is fixed. What changes in relation to what, can be answered finally only in hypothetical terms by reference to some inertial frame of reference as formulated by Minkowski. But how one can establish that one is a member of this hypothetical special class of privileged observers and of objects all of whom and which are at rest or in uniform relative motion is impossible to say. Our practical purposes will be better served by recognizing that everything is changing, and by giving up clinging to the false security of believing that things can remain in *exactly* the same state or at *exactly* the same place. Identity means continuity of existence; it does not mean static constancy and exactitude in form and place, any more than it means constancy in time.

Measurement and the Definition of Time

Finally, we may sharpen and extend our concept of time by defining its length operationally, by reference, that is, to the operations required for the measurement of length of time. We shall consider this matter in a preliminary way at this point, and shall return in Chapters 10 and 11 to certain more general questions of the consequences of time measurement for the social and psychological sciences.

The measurement of length of time requires the discovery and fixing of equal intervals between points in an event or change process, that is to say, equal intervals between temporal points—just as the measurement of length of space requires the discovery and fixing of equal intervals between momentarily related points. The history of measurement of lengths of time may be encapsulated in three main phases: the use of calendars and divisions of the solar day, sundials, and sand and water timers; the use of mechanical clocks since the fourteenth century, with the discovery and improvement of the escapement mechanism and its combination with the pendulum; and the contemporary development of exceedingly accurate atomic timepieces (currently the cesium clock) which so far as is known do not suffer from variations in movements of the earth or, indeed, its gradual slowing down.

It is not necessary here to pursue in any greater detail this most fascinating technical history.[13] But there is one element which must be teased out and identified, namely, that the measurement of time is in fact the measurement of *length* of time, distance in time, distance between two temporal points, and it requires that "time be stopped." That is to say, in creating units of time we have to create an intellectual artefact, a conceptual construction, one in which processes like the movement of a clock stand still.

This process of making the movements by which we measure time apparently stand still is a consequence of all conscious objectification of thought by its transformation into shareable language. It holds in the same way with respect to space: spatial points have to be artificially stopped, to be made to stand still[14] while we measure the distance between two momentarily simultaneous points. For some reason we refer to this distance as "length" rather than as length of space, in contrast to the distance between two temporal points which we always refer to in full as "length of time." I believe that this semantic usage derives from the primacy with which we endow space on emotional grounds because of our felt familiarity with it and its seeming stability, as against time which seems inaccessible and shifting.

The measurement of length of time, therefore, calls for the fixing of units which can be carved in thinner or thicker slices out of processes. These slices must be able to be frozen conceptually and written down (recorded). And slices of conceptually equal thickness must be able to be constructed and identified. Examples of such slices are the year, the month, the day, the sundial hour, the sand- or water-flow hour (or shorter unit), the swing of a pendulum unit, the mechanical escapement unit, the atomic cycle unit. Each of these mechanisms allows for the identification

[13] See, for example, the excellent description by H. Allan Lloyd, (1966), "Timekeepers—an historical sketch," pp. 388–400.

[14] Although we ordinarily do not notice this fact because we tend to see most things in space, falsely, as standing still anyhow—as we argued in the case of the paradox of the same thing at the same place at two different times. (See above.)

of a boundary—a temporal point—between successive phases in a process; and for different reasons in each case, allows for the assumption of equal intervals between these boundaries or temporal points.

It is this last point which has often made the foundation of time measurement (or length of time or distance between temporal points) seem less secure than that of the measurement of space (or length of space or distance between spatial points). We seem to feel that once we construct a standard spatial-length rule, its divisions remain constant. By contrast, we seem to feel unsure that the divisions on our clocks, for example, remain equal, "because we cannot compare them directly with one another at different times." Again, the perceived difference is an emotional distortion. Two different spatial-length yardsticks can, of course, be compared directly with each other: but so also can two different clocks. But the length of intervals on the same clock cannot be compared directly with one another at two different times: but neither can the length of an interval on the same yardstick be compared directly with itself at two different times. Once again we readily accept spatial constancy but not temporal constancy; the recognition in relativity theory that there occurs a change in the length both of spatial yardsticks and of temporal yardsticks with change in velocity should have destroyed this outlook.

Temporal yardsticks, therefore, allow us to abstract episodes or events from the continuous succession of perceived change and transformation in the world around us, and to order these experiences to objective clock units. We can by this process get objective measurement of the length of time or duration of objective processes—in the terms in which we defined the concept of objectivity above. The method of constructing clocks, and of ordering the duration of episodes or events to those clocks, constitutes a definition of the distance between objectively established temporal points, that is to say, a definition of length of time. This definition (of length of time) extends the general definition of time already given; it is no less objective than the definition of length of space.

Externally objective clocks, like externally objective yardsticks, can also be referred to in the internal world of thoughts, fantasies, and dreams. But because these processes are themselves internal, the measuring process cannot be externally shared: it remains imaginary, but in our terms nonetheless "real" for all that. That is to say, they are mentally real, but in order to become objective the results of the measurement must first become subjectively formulated, and then be communicated to others. This distinction, to which I have already drawn attention several times, will become of central importance when we turn to consider the conception of two axes of time and the consequences of this conception for the understanding of human behavior.

PART TWO

THE EXPERIENCE THAT IS TIME

CHAPTER FOUR

The Conscious, Preconscious, and Unconscious Experience Called Time

I have so far considered the case for the reality of time, not as a thing-in-itself, a *zeit-an-sich,* but as a concept abstracted and constructed from the experience of succession, of process, of events, of continuity, in our total world; that is to say, from the experience of the occurrence of events in both the external material world and the internal psychological world. The concept of time so constructed is analogous to the concept of space. It is a univocal concept, applying to all types of experience, ponderable and imponderable, objective and subjective, and to experience in thought, imagination, and dreams.

One difficulty, however, is that we tend to reify time, confusing the concept, which is a static idea (as are all concepts in and of themselves) with the experiences gathered together (*con-ceptus*) within the concept. It is thus that we incorrectly speak of the flow of time, as though time were the term for something like a river—as for a concrete particular, rather than for a positional noun or a universal category. This reification of time by particularizing it in concrete terms contrasts with our attitude toward space. We do not treat of space as though it were a physical thing, a box, which stays still while we stand in it (not since Leibniz, that is), and when we now refer to the curvature of space we are aware that that is a mathematical construction and not somehow an infinite plasticine ball that has been pressed out of shape.

I shall consider this reification of time, and will show that in raw experience it is our sense of a space–time manifold, or plenum, which suffuses our awareness, and not our sense either of space or of time by itself. In considering this matter I shall suggest that we must recognize three different levels of component which together make up the totality of experience, and which are enfolded within our idea of time—conscious experience, preconscious experience, and unconscious (protomental) experience. I think that the theory of knowledge in general is hamstrung by its failure to separate out these three different components of mental functioning. This shortcoming is especially pronounced, however, in the case of the conceptualization of time phenomena.

I shall argue that it is the failure to recognize the existence of these interlacing components of experience which gives rise to such questions, puzzles, and paradoxes as: does time flow? does it have direction? is there a past as well as a future? is there merely earlier and later? do we live in an atomic unchanging world or in a world of flux and durée? For each of the three elements in experience—conscious, preconscious, unconscious—creates its own characteristic picture of the world, and the answer to each of these questions depends upon which element of experience is assumed: the more conscious, the more discontinuous and static; the more unconscious, the more continuous and the more in directional flux.

Finally, I shall have to establish that reason, logic, and rationality[1] are not solely the prerogative of the conscious mind, of conscious mental activity. They are always the outcome of the interplay between conscious, preconscious, and unconscious mental activity in this sense: conscious mental activity sets the explicit articulated framework of behavior, including the context of knowledge within which we act; preconscious mental activity provides the background store of knowledge and awareness upon which we can consciously draw; and unconscious protomental activity provides the continuously shifting direction of intentionality, the sense of where we want to go, wish to go, will to go. This distinction contrasts sharply with more common usage which takes it that it is in the conscious processes that we find the reasonable, the logical, and the rational, as compared with unconscious mental activities which (if they are granted existence at all) are regarded as the seat of the illogical, of unreason, and of the irrational, or at best as the source of foolhardy rationalizations.

My reason for bringing into my analysis these questions of the logical and the rational in relation to the conscious and the unconscious, is to pursue the following argument. Unconscious desires and goals are elements in rationality. Goals are intentions; therefore intentions can derive from unconscious protomental sources and at the same time contribute

[1] I make the distinction between: logical, as degree of adherence to formal rules of thought; reasonable, as judgment that action is broadly in accord with intent; and rational, as extent to which activity is reality-based. It is useful to note that etymologically all three terms reveal their roots in both conscious knowledge and in unconscious sensing: logic from Gr. *legein*, to collect, gather, and select, as well as to tell and to speak; and reason, and rational, both from L. *reor*, which is not only to think, but also to judge, and to deem.

to rational action. Our experience of goals and intentions underlies the meaning of the future, just as our experience of memories underlies the meaning of the past. Our predictions or ideas of the future, and therefore of the so-called directionality of time and of flux, can be understood as the expression of the conscious mode of formulation of unconscious states of mind in the present with their ongoing desires, goals, and intentions.

The Reification of Time

As I have indicated in the previous chapter, much of the debate about the nature of time is a fruitless debate, arising from the reification of time. We often treat it as a concrete thing. If we did not, we should not get into arguments about the passage of time, or the flow of time, or about the future flowing into the present into the past, or about whether the arrow of time is unidirectional or bidirectional or directional at all, or whether there is a time arrow, or about the possible effects of entropy upon time. These formulations confuse the concept of time as a positional noun and as a universal category, with the experience of the concretely particular events or processes involving concrete particular things, which are the phenomena from which we construct the concept of time.

It can be stated quite simply that time is neither like a river nor like an arrow. *Tempus fugit* is literally not true. Time does not either flow or fly—regardless of such commonplace expressions, when people are engrossed in what they are doing, as "how time flies," or "how time has flown," or "how quickly time has passed." What they mean is that this particular event has taken a greater number of minutes or hours than they had anticipated or realized. Nor does time go in any particular direction, because time *per se* does not go anywhere or point anywhere.[2]

This reification of time is similar to the reification of space which characterized thinking at the time of Newton. Newton himself employed a static concept of space, not as an ordering of the material world but as a substantial entity which was a container of objects and which thus was assumed to have existence independently of the existence of any material objects. It was this view which was argued against by Leibniz, who saw space as a matter of the relative positioning of objects, and shortly after by Kant for whom space was an *a priori* category.

The view of Leibniz has been the dominant view of space in modern

[2] The concept of time is in this respect no different from any other universal. Universals do not *have* qualities or properties; they *are* qualities or properties. No valid propositions can be made about a universal other than a statement of its existence. For example: the proposition "Man is a category comprising all men" is valid; but "Man is alive" or "Man breathes" are invalid—they require to be formulated as "That man is alive" or "All living men breathe." Or, to take another example, "Red is a color category" is valid; but "Red is shocking" is invalid, and requires to be reformulated as "All red colors are shocking" or perhaps "Some red colors are shocking."

science, especially in relativity theory. The concept of space has been abstracted and generalized into a mathematical construct. This construct can be subject to any type of mathematical manipulation depending upon the hypothesis used. It can be stretched and curved into any shape—any mathematical shape, that is—and straightened out again, but without our having to picture some kind of physical entity actually changing shape.

With time, however, we behave otherwise. McTaggart's argument that the future cannot coexist with past and present, and that therefore time cannot exist, is an argument that is based on the presupposition that a time–thing either exists or does not exist, rather than a discussion of the possible usefulness of a positional time–category. And even Popper, in his protests against the idea of the reversibility of the direction of time (because it would mean, for example, that the fact that Hiroshima had been perpetrated by man would be reversed and would disappear from human awareness), is at the same time arguing that there is a particular thing called time and that this thing does move, albeit in one direction only. Similarly, the arguments of Lucas[3] and Dummett,[4] to take but two examples, seem to take for granted that there is a moving thing called time—the argument being an argument about its properties.

But what if it should be desired to avoid the reification of time?[5] What is the alternative? The alternative is to treat it as a positional noun or as a universal, rather than as a pointing term referring to a particular identifiable thing or phenomenon. In doing so, we are forced to ask ourselves what kinds of data, what kinds of experience, what kinds of phenomena, are to be subsumed under the concept. And it is at this point that the questions become interesting. For we are forced to ask ourselves in what way do we experience the past; or the present; or the future? Just what seems to have speeded up? Or what do we mean when we say that time's arrow or the passage of time seems to be in one direction?

In order to try to deal with these questions I propose to consider the idea of time from the point of view of our conscious experience, our preconscious experience, and our unconscious experience. I believe it is essential to carry out this threefold analysis, since the phenomena associated with time differ for each of the modes of experience. Conscious experience gives us the focused verbalizable perception of things moving or changing, of events; but this focused perception is organized into a

[3] J. R. Lucas, (1973), *A Treatise on Time and Space*.

[4] Michael Dummett, (1954), "Can an Effect Precede its Cause?"; and (1964), "Bringing About the Past."

[5] It would perhaps be better to have spoken of spatiality and temporality, rather than of space and time—but this awkward usage is unnecessary so long as we keep in mind what we mean when we say, "What time is it?" (a particular reading on a clock), as against "Time, what is it?" (temporality as a universal).

static, discontinuous, atomic world in which time phenomena are dominated by the spatialized notion of discontinuous ticks on a clock.

Preconscious experience gives the background or surround to our flitting consciously focused and verbalized percepts. It is the peripheral awareness in which we sense the ongoingness of the things we perceive as in motion, and by means of which we can formulate our sense of the extension of events in terms of continuous flux and dureé.

By contrast, unconscious protomental experience is unverbalized. It comprises the psychological world of the continual flow of desires, of passion, of goals and intentions and will. It is the world of primally fused memory, perception, desire, and intention (the unified field which exists before we consciously differentiate the separate parts), combined into what might be termed the moving present, a present which is felt as moving from out of the past and into the future. It is these unconscious phenomena which give us the notion of time as having a direction which expresses the goal-directedness of intentional behavior.

I now propose to consider this threefold categorization of experience—for which, of course, Freud was responsible—more fully in connection with its significance for the clarification of the full complexity of the human idea of time. It will take us into a formulation of a world which is neither atomic nor in flux, but both; neither continuous nor discontinuous, but both; made up neither of forces acting between points at a distance nor of fields of force, but of both; neither static nor flowing, but both; with neither past—present—future nor earlier and later, but both; neither predictable nor retrodictable, but both; neither universal nor particular, but both; neither objective nor subjective, but both; neither material nor ideal, but both; neither concrete nor abstract, but both.

This formulation is that of a phenomenal world. But it is not the limited phenomenal world of Husserl, built upon the unnecessarily narrow idea of mental life as exclusively conscious. It is a world of kaleidoscopic interaction of conscious, preconscious, and unconscious phenomena—with a continual restless oscillation between the dominance of one mode then another. It is a world still at one moment, alive and in motion at the next; vague and cloudy, then clear, external and atomic; full of hazily felt intuitive introspect, then acutely organized with known external real objects; an inner world, then an outer world; a world of just-out-of-focus periphery and a world of sharply focused things; a world of passion, feeling, and desire, then of intellectually grasped impassive idea; a world of love and anger, and then of mathematics and cerebral logic; and most of all, a world underpinned by unconscious processes which are simply inaccessible to conscious introspection, processes which can be sensed but not observed, and which disappear under the impact of verbal formulations since their essence is orientation and action and not words.

These varied facts of conscious, preconscious, and unconscious processes are the means by which we construct and experience our world. They are the means by which we act upon our world and react to it, and by which we build up our conception of it—including not just the conception we call time, but also space and space–time, number, quality, form, extension, whole and part, and chair and man and all the other categories we construct from our constitutional endowment playing upon our experience and the way we organize it. Any theory of knowledge must take all these three levels of mental process into account. I shall do so in the following description of the phenomenal roots of the concept of time.

In order to formulate my propositions I shall employ these terms: *conscious knowledge; preconscious awareness;* and *unconscious sensing.*

Conscious Knowledge and Time: Focused Verbalization

If we consider the flow of our experience carefully and in detail, it may be observed that it is neither a wholly continuous nor a wholly discontinuous one. Our conscious attention flits rapidly from thing to thing, and from idea to idea, and what is in focus at one moment falls into the background the next.[6] But underneath this restless activity there is a continuous flow of unconscious sensing, intuition, hunch, judgment—a sensing which regularly becomes a diffused field of attention replacing the sharp conscious focus, a diffused field which is extremely difficult to get one's hands on and verbalize.

The recognition and identification of these continual shiftings of ideas between conscious focus and diffuse background is a central feature of psychoanalytic theory, and led Freud to his division of mental systems into conscious, preconscious, and unconscious.[7] Consideration of the nature of the experience of time within each of these mental categories

[6] This discontinuous process may be observed in the following way. If you stare hard at a blank wall, or a printed page, you will observe a rapid oscillation—occurring in fractions of a second—in which the wall or the page is in focus, then disappears for an instant while awareness seems to turn inside, then reappears, then disappears again, and so on. I believe that this inward-turning of attention occurs, for example, in the extremely rapid movement phases of eye movements in reading. It is as though we are constantly maintaining our adjustment between our external and internal worlds by this continual perceptual oscillation between outside and inside.

[7] Sigmund Freud, (1923), *The Ego and the Id*. Sensitive to the criticisms of philosophers about the idea of *unconscious* mental functioning, he wrote (p. 13): "To most people who have been educated in philosophy the idea of anything psychical which is not also conscious is so inconceivable that it seems to them absurd and refutable simply by logic. . . . Their psychology of consciousness is incapable of solving the problems of dreams and hypnosis." To which I would add, "and certainly incapable of solving the problem of time."

will help to resolve the apparent conflicts between the atomistic view of time, time as flux, and time as future intent.

Let me illustrate by the following situation. I am holding a pen with which I am writing these words. If I attend to its movement across the page I may be vaguely aware of its permanent qualities but cannot focus upon them. If I focus upon its being an object—its "thingness"—then its motion clicks out of focus and it seems for that moment to be still. If I note its black color, then I may be vaguely aware of its motion but I cannot at that moment focus upon it. If I note its tapering shape, color and motion disappear into the background. If I listen to its quiet gliding sound as it writes, motion, color, shape, and all other qualities are held perceptually somewhere else, temporarily inhibited and held out of the center of attention.[8] Moreover, as Schachtel[9] has shown, the process of paying attention by conscious focusing is as much a matter of discriminating by inhibiting those parts of the field you do not wish to perceive so as to leave only that region in the field upon which you wish to focus, as it is of positively focusing upon something. In order to focus upon A we inhibit, block out, push to the background everything which is not-A, so that only the region A is left.

Why should it be, however, that there are certain modalities which stand out by themselves as self-contained, formed, segregated wholes, on a take-it-or-leave-it basis? The discreteness of certain perceptual modalities might seem natural in this regard—that is to say, it does not seem unreasonable that one should be able to focus separately upon the color, or the sound, or the smell of pouring coffee, but not all at the same time, since these belong to identifiably different sensory processes. But what about the categories of shape, number, size, movement, constancy—the categories of state or position—which are not connected with any one particular sensory modality? Why should we see the world in terms of ideas of this kind rather than of some other kinds of ideas—kinds which it would be difficult if not impossible to imagine?

It is these categories of state or position which have been explored by Gestalt psychology. They arrive on the scene whole and in their own right, even though they may require a particular culture in order to emerge. Their potential to emerge as Gestalten is, *a priori,* constitutional, innately given, and not built up out of experience by the association of simpler

[8] In *The Ego and the Id* (1923), for example, Freud writes (pp. 13 and 14): " 'Being conscious' is in the first place a purely descriptive term, resting on perception of the most immediate and certain character. Experience goes on to show that a psychical element (for instance, an idea) is not as a rule conscious for a protracted length of time. On the contrary, a state of consciousness is characteristically very transitory; an idea that is conscious now is no longer so a moment later, although it can become so under certain conditions that are easily brought about."

[9] Ernest Schachtel, (1963), *Metamorphosis*.

elements.[10] They constitute natural contents and natural lines of cleavage of experience. They include the experiences with which we have been most concerned: constancy, continuity, distance, motion, and succession, and our sense of space and time. They also cannot be in focus at one and the same time.

If we now turn to time, and focus upon the motion of an object, our conscious awareness of its substantial permanence disappears, and vice versa. If I focus upon how long my pen takes to move across the page, I cannot simultaneously focus upon how long my pen is in length (in spatial length, that is). I cannot thus focus at the same time upon the direct percepts conceptualized as space and upon those conceptualized as time. If I focus upon the immediate spatial distance between two different points, then the length of time of processes falls into the background, along with shape, form, sound, color, and a host of other modalities. And if I focus upon the temporal distance between two points in a process, then the focused perception of immediate spatial distances disappears into the background.

These oscillatory processes are familiar enough in the figure-ground perception experiments, as in, for example, the well-known vase and face figure shown. When we see the faces as figure, the vase has become ground, and vice versa; they cannot be seen together in focus. Our experience of the permanence and succession associated with space and time is the same: we cannot experience them in conscious focus together. How then can we expect to experience the space—time manifold? The fact is that we cannot do so. Not in conscious focus at least. Consciously

Figure 4.1. Reversing face-vase figure.

[10] It is not necessary to review the voluminous literature in which the existence of *a priori* Gestalten has been experimentally demonstrated not only for perception but for more complex psychological processes as well. See, for example, Wolfgang Kohler, (1929), *Gestalt Psychology;* Kurt Koffka, (1928), *The Growth of the Mind,* and (1935), *Principles of Gestalt Psychology;* Kurt Lewin, (1935), *A Dynamic Theory of Personality.*

we can experience not a oneness but only a summation of simultaneous points at a distance (giving space) and successive points in a process (giving time) by rapid oscillation of attention between formulated conscious knowledge of the two. In this same way, we could consciously experience our drawing as a face—vase manifold only as a retrospective conceptual construction, an intellectual addition of the two aspects.

But what then do these consciously focused mental processes have to do with our conception of time? I would draw attention to the following features. These focused processes are experienced as discrete percepts such as "that pen I see," or "that sound I hear," or more simply perhaps "that thing or object I see"; or else as delineated ideas, such as "that movement I see," or "that melody I am listening to," or "that distance between pen and paper."

The fact of the discreteness, and the putting into words, both contribute to the atomistic view of the world which derives from conscious focused experience. The conscious world is a world of objects, of things. It is a static world. It is a mechanical world, a world in which forces act between objects at a distance from one another. It is dominantly a spatialized world, a world characterized by discontinuity, a world which is describable in terms of geometry and arithmetic and mechanics. The nearest this world gets to continuous flow is in infinite series, with an infinite number of beads strung along a thread at an infinitely small distance from one another.

These discontinuous and spatialized qualities of conscious knowledge lead to the atomistic conception of time. This conception arises from our conscious perception of events as a discontinuous change in position of objects. Thus, for example, the movement of a car is perceived in the conscious perceptual mode in terms of a retrospective construction: "the car has moved from A to B in three seconds" means that three seconds ago I saw this car for a motionless moment at point A, and now I see it for an equally motionless moment at point B three seconds later.

It can thus be seen that our conscious knowledge of time is the time of earlier and later. It is the clock time of the physicist, the time of a succession of later and later temporal points. There is no flux, no durée, in this world. Taken by itself, the conscious knowledge of time conforms to the Parmenidean and Platonic view that all is constant; there is no change. It leads also to the famous paradoxes of Zeno: so long as all is conscious, then all we can have is a series of points of time which, even though infinite in number, nevertheless can never move. The arrow is forever still—at least so long as we believe that all of experience is conscious, that there is nothing more to mental life than conscious mental life.

Preconscious Awareness and Time: The Focusable Context

But the fact is that there is more—indeed very much more—to mental life than mere focused conscious perception and knowledge. The conscious components of experience, the knowledge, are certainly important elements of experience, but they are just as certainly no more important than either the preconscious or the unconscious elements. I shall now consider the general characteristics of the preconscious background phenomena as compared with the focused foreground, and then consider the unconscious in its own right—both, of course, with particular reference to the experience contained within time.

The notion of preconscious mental functioning was formulated by Freud. As he put it, an idea may be in conscious focus one moment, then not in conscious focus, and then it might be brought back into focus again, but "in the interval the idea was—we do not know what. We can say that it was *latent,* and by this we mean that it was capable of *becoming conscious* at any moment."[11] Then he adds, "The latent, which is unconscious only descriptively, not in the dynamic sense, we call *preconscious.*"[12]

The experience of preconscious background is a difficult human experience to come to grips with and to formulate. It is difficult to formulate precisely because it is background and therefore not structured and focused as is the foreground. To formulate is to give organized form to something, and at that point it is ready to be put into words. But if the nonformed is formulated, made into segregated wholes, it loses its nonformed structure and the preconscious becomes conscious—that is what always happens with words.

The fact that the preconscious nonfocused, nonformed, nonsegregated, nondiscrete wholes type of experience is difficult to put into words is no new idea—to put into scientifically rigorous words, that is. It is better expressed in poetic language, where the sense and color and tone and sequence and emotional resonance of the language can perhaps mirror and portray the quality of the experience.

Let me illustrate what I mean, by the following example. Hunters know that moving objects are most readily detected by peripheral vision. To use this ability requires that one's vision be focused ahead, but also that somehow one keeps a sensitive awareness of what is going on round about to the side, on the periphery. If something is seen to move on the periphery, the eyes can then be directed toward the spot. But how can one describe what one "sees" peripherally, preconsciously? It is not

[11] Freud, *The Ego and the Id,* p. 13.
[12] Ibid., p. 15.

possible to do so. There is what Köhler refers to as the comparative "emptiness" and "looseness" to the ground, whereas the figure has "the substantiality of a thing. . . a character of solidity and coherence."[13]

But nevertheless we have an awareness of the periphery as well as of the focus of experience, of ground as well as of figure, of preconscious as well as focused conscious, a sense of things going on to the side of perception, around its circumference somehow. It is part of our reality, but it has the quality of a continuous field of regions which are not sharply delineated from one another. The whole area is flat and colorless and boundless in extent, and the field "swims" like a sky of flat moving clouds. In this field space, time, form, and all the other modalities are there together in a state of fused Being and Becoming. The words are impossible to find, for the moment we speak or write of space, time, form, etc., even if we say they are fused in a field, the mind nevertheless follows the separate words to separate entities, regardless of our statement that they are fused. Consciousness reaches into the preconscious at every opportunity, fixes upon particular areas, brings them out into conscious focus, and atomizes them—objectifies them—verbalizes them.[14] And then, for the moment, they are no longer preconscious but conscious.

The preconscious awareness of ground, or context, then, is unfocused, it does not contain things, it is a continuously extended field, never-ending, and it is constantly in flux, shifting and changing in pattern and configuration. In contrast to consciously focused perception which gives momentarily stilled snapshots, preconscious awareness gives a direct experience of movement, of motion—it is like a motion picture but without the focused picture. It is this direct awareness of motion or flux which underlies the meaning of time as flux or durée rather than simply as spatialized points. It is the meaning of time formulated by Heraclitus and by Bergson.

It is in preconscious experience also that we get our direct experience of the space–time manifold; not space and time, but an awareness of a world of spatially extended process, of space and movement in space all intertwined. The space–time manifold *per se* as a context or ground phenomenon is a description of an aspect of the experience of the background of things, but not in itself perceivable as a thing in focus as figure. The fact that terms like space—time manifold, or space—time field, or

[13] Wolfgang Köhler, *Gestalt Psychology,* p. 219.

[14] Freud has observed, "The question 'How does a thing become conscious?' would thus be more advantageously stated: 'How does a thing become preconscious?' And the answer would be: 'Through becoming connected with the word-presentations corresponding to it.' " Op. cit., p. 20.

space–time plenum are used to refer to it is not accidental. These terms are very general terms which refer to a sense of unformed fullness.[15]

The analysis thus far, therefore, leads us to two of the different modes of describing time—as of everything else. One is the conscious atomistic concept, related to changing things in the focused foreground or figure. The other is the preconscious field concept, related to the awareness of continuous flux in the contextual field or background of experience, in which time is embedded in the space–time manifold. Neither is the more true concept of time. Both are necessary—just as atomic theory and field theory are both necessary in physics and, if our analysis is correct, always will be necessary, not as an unsatisfactory and temporary intermediate situation as Einstein thought but as a fact of the structuring of all human mental activity into figure and ground.

Unlike Minkowski, therefore, I would not want to fuse space and time together forevermore and never treat them apart. I shall argue for the retention of both modes of description—the retention of both Bergsonian durée and Heraclitean flux, on the one hand, and the physical time and atomism of Democritus, on the other hand. Meanwhile, we may note in passing that it is theories of durée and flux which are most difficult to put into words, which are most likely to seem unscientific and even mystical, and which constitute the particular meaning of time which Hegel described as killed by words. This difficulty will always exist: discontinuous words cannot accurately map continuous process.

The Sense of Temporal Directionality

Conscious figure and preconscious ground constitute a dynamic whole with a continual restless movement in which ideas or objects pop into focus and then sink back again into the blur of the unfocused ground. With respect to time, this conscious–preconscious interplay has yielded us the immediate experiences which lie at the roots of the atomistic clock time of dating and of earlier and later, and of the time of fluxion and durée. But neither consciously experienced clock time nor preconsciously experienced flux has yielded the direct experience underlying our sense of temporal directionality: spatial directionality, yes; but temporal directionality, no. This point is a subtle one, but extremely important. Let me try to clarify it.

In order to look at directionality, let us consider the immediate

[15] Manifold refers to the numerous in a general sense; it was used by Kant to describe the sum of everything sensed *before* being organized by the understanding. A plenum is the opposite of a vacuum: it refers to a totally filled space. Neither of these terms describes focused perception. It is striking also that the general meaning of a field is ground or extended surface, a field as a limited and segregated piece of land being a special usage.

experience of spatial direction first. A sense of direction of processes or events in space can come about by two means. First, it can be experienced immediately by the focused perception of an arrow, or of any other pointing sign or symbol, or by the background of awareness of the move-

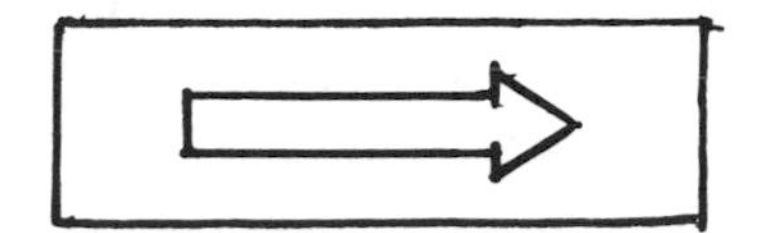

Figure 4.2

ment of an object in a particular direction. Second, it can be experienced as an immediate idea by the reconstruction of a movement; for example, in the idea that an arrow which is embedded in a target can be preconsciously experienced as having moved from the bow from which it was shot, in the direction of the target to which it flew.

When we turn, however, to directionality in time, the flight of an arrow and spatial directionality do not help us. For directionality in time runs from future to present to past, and its direct experience calls for an immediate sense of the coexistence of past, present, and future. But it is no use to say that we can somehow have an immediate experience of an arrow flying out of the future into the past. Such a description would be suitable as a poetic metaphor giving life to an arrow by attributing purpose to it. But recourse to magical anthropomorphism of this kind will most decidedly hold up our argument rather than carry it forward.

Our direct temporal sense of the flight of an arrow, or of any other external movement, is a sense only of earlier and later. To say that the arrow is moving from the bow toward the target is a statement of its direction or sense in space; it is not a statement of its direction in time. We can directly perceive the spatial direction of a movement as something going from "there to there"—all in the present. But we cannot get the same direct sense of past to future in the perception of the movement, precisely because it is all in the present. The arrow can be said to have left the bow at a time t_1, five seconds earlier than it reached the target at a time t_2—so long as we recorded both points in time on a stopwatch. But these earlier and later recordings of time are static: they do not in any way give a direct and immediate sense of future—present—past movement, of future—present—past direction. The bow, the arrow, the target, all exist as spatial points, having moved between two different locations at the two different time recordings, and can so be experienced. But they do not move in any particular direction temporally.

Where then does our sense of temporal direction come from? That is to say, where does our sense of flux from future, to present, to past, come from? To say that A occurred earlier or later than B is a matter of objective record (or at least of objectifiable record): it can be recorded in the external world. When we shift to past, present, and future, however, we would appear to move into the inner psychic world—the world in which past, present, and future all coexist, as Augustine noted, as the present past, the present present, and the present future. They coexist, that is, as the interacting field of present memory, present perception, and present desire or expectation and intention.

I would now draw attention to a very significant circumstance. We have a sense of the coexistence and interaction of past, present, and future because memory, perception, and desire exist as a unified field of force. It is from this unified field of force that we get our sense of temporal direction just as surely as a pointing arrow gives a sense of spatial direction. But the arrow of temporality cannot be physically observed; it is not like a physical arrow somehow pointing from a later time than the present to an earlier time than the present. It is a peculiarly psychological directionality—a peculiarly psychological arrow. It is a mentally felt pointing, an orientation of the mind, sloping or slanting, from what might be thought of as a future time in the field which constitutes the present to a past time in the same present field.

Thus, for example, I note that I am thirsty and get a glass of orange juice from the refrigerator. Such an action comes about through the fused interaction between: a sensed need, a lack, the thirst; a seeking for an idea of something which might satisfy the need; a memory of water in the tap, but then of orange juice in the refrigerator; a playing together of the sensed need, the idea of the remembered orange juice, the sense of what would need to be done to get and drink the orange juice, and the potential satisfaction in doing so; the construction of the intention to get the orange juice; and the carrying along of the memory, need, and intention in the willful action of getting to the refrigerator, pouring the drink, and quenching the thirst. In this example, temporal directionality can be experienced directly in the simultaneous experiencing of the thirst and the intention to quench it in just a few moments.

What then are we to make of this psychological field of force in which past, present, and future rub shoulders with one another, jostle one another, walk together hand-in-hand, work together, cooperate with one another and pull against one another? A moment's reflection suggests that we are here dealing with living experience, with the present moment in full regalia in all its richness. Let me examine the present past, the present present, and the present future, each in turn.

The interacting past is a highly selected past. It is the past of immediately relevant memory. Particular memories are somehow chosen and drawn from the great storehouse of memory—or somehow extrude themselves—to lend the current meaningfulness of the precipitates of earlier experiences to the handling of the present situation. These memories are alive in the present. How their meaning relates to their meaning at the earlier time of their occurrence can never accurately be known. But they inform the present with their accumulation of wisdom and knowledge (and of errors as well), and indeed sometimes press for particular outcomes as much to satisfy continuing frustrations, leftovers from earlier experience, as to satisfy current need.

If we next move to the organizing present we find an amalgam of current percepts—responses to selected outside circumstances—and current lacks or needs or interests, the things which are astir in the person, and which cause attention to be directed in a manner calculated to search out the most useful information or resources—both from memory and from the outside world. The existence of the orange juice in the refrigerator is winkled out of memory, or presses out of memory, through the interaction of memory and of perceived need.

And then, finally, there is the future, the seemingly nonexisting will-o'-the-wisp. It is the product of our experience of the things we intend to do, to achieve, to create, to bring about. These things are the goals or objectives which we have in mind, which when achieved—or so we believe—will provide us with what we need to satisfy our felt lack, our desires, our expectations. The future is that which we plan to do to express our current interests, to satisfy a current lack, or to bring about what we have decided we want. The future is a statement not of some actual event which has still to get here, but of our will or intention—whether to get a drink, as in our example, or to paint a picture or a house, or to take a holiday, or to create an export sales network, or to preach a sermon, or to land a man on the moon.

Present needs or interests express themselves in selective attention or perception, and arouse memories which in turn influence perception, and lead to the imagining of things-which-might-be. Remembered things and ideas merge with perceived things and ideas, and lead to desired things and ideas, and all fuse in one active field of force generating willed action toward an intended goal. That is what life is about.

Temporal direction, then—the directionality from past to present to future—is nothing more nor less than the fusion of experience, of anticipation, of need and perception, and of memory, into a single force field in action toward an intended, desired, or willed goal. It is the sensing of the goal-directedness of our personal endeavors and behavior.

If this view is correct, that the essence of temporal directionality lies in a person's intra-psychic sense of fused memory, perception, desire, and intention, where then in mental life might such a fused experience—a psychological field of force—occur? Certainly not in the conscious region of mental activity. For there can be no spontaneous conscious knowledge of a memory—perception—desire, or a past—present—future, all together as one focused phenomenon, as a field of force, any more than there can be a focused conscious knowledge of a space—time plenum or field.

We have seen how consciousness is the great abstractor, the great divider, the great still-photographer. It creates our knowledge in terms of experience of discrete points at a distance and not of fibrillating fields of force. That characteristic is equally true whether it is conscious perception of the external world as objects, or the internal world as ideas. The ideas too—whether consciously repeated, or drawn from the preconscious, or sliding or erupting from the region of the unconscious—are produced in knowable and verbally formulatable bits. I may construct a conceptual model of a field of force, and consciously know that model as a thing. But that is to know the field of force only at one remove, at secondhand. Firsthand knowledge of a field of force by direct experience is not available to conscious mental processes.

Nor can fused past—present—future be found in preconscious awareness. Preconscious mental processes tend to be passive background containing a store of focusable and verbalizable elements. These processes do not generate intentionality, they do not generate goals. Hence they cannot generate the temporal directionality which derives from the sense we have of pressing toward a goal in time.

Unconscious Sensing and Time: Protomental Source of Temporal Direction

If not in consciousness, and if not in preconsciousness, then where is the dynamic field, the field of force, comprising the coexisting past—present—future or actively interpenetrating memory, perception, and desire, to be found? I emphasize again that I would speak not of each separate element by itself, nor of the intellectual verbal formulation of the field, but of the direct experience, the ongoing active sense of the field, the gut feel of the process. This powerful flow of intentionality is the outstanding characteristic of unconscious mental activity.

For the world of figure and ground, of the conscious—preconscious system, does not operate on its own. It is only the surface part of mental activity. Beneath this surface lie the vastnesses of the unconscious mind—

the protomental region —in which the bulk of mental activity proceeds.[16] It is in the reaches of the unconscious protomental activity that are to be found the dynamics of human action, the reasons for our doing things, the source of our desires, goals and objectives, and our decisions. It may be clear that I am here using Freud's full concept of the unconscious system—the unconscious as a part of normal mental functioning as well as the system into which previously conscious or preconscious ideas can be repressed and split off with resulting abnormality in function and symptom formation. For the moment, I shall be concerned with normal unconscious functioning and its connection with time, and shall turn to the questions of psychopathology and abnormal temporal phenomena in Chapter 12.[17]

In considering protomental unconscious life we encounter a problem which if not identified and got out of the way will gravely obscure the path I propose to follow. It is the fact that in writing about the unconscious, putting it into words, the essence of the normal unconscious (as against repressed and split-off unconscious processes), its fluid totally nonverbal character, is lost. And so we risk losing the idea altogether.

We encountered similar difficulties in describing preconscious processes because they are a continuous field and are therefore ill-dealt with by words which are by their very nature atomistic. But these difficulties are much greater with respect to the unconscious field. Preconscious processes are at least available in latent verbalized form: it does not, therefore, totally macerate the preconscious processes to describe them verbally. Unconscious processes are different. Except for the repressed and split-off unconscious elements which must previously have been conscious or preconscious and verbalized, unconscious protomental experience cannot in any way be adequately verbalized. To describe it in words is the same as describing in words the way in which a two-week-old baby is experiencing his world—or an animal experiencing its world. It is not impossible: but it must be constantly kept in mind that the *verbal description is a description of a nonverbal process* and not a recording of words used in the unconscious. Unconscious processes are non-artic-

[16] Tolman described the matter as follows, in connection with animal learning: "The rat must have in some measure learned unconsciously, before consciousness can begin to appear. He cannot be conscious until he has already achieved to some degree, that which being conscious will merely further emphasize. The functional value of being conscious must be said, therefore, to be merely that of emphasizing and reinforcing field-features, to some extent already becoming immanent in preceding behaviors. *Consciousness can achieve nothing really de novo.*" [Italics mine.] E. C. Tolman, (1967), *Purposive Behavior in Animals and Men:* p. 208.

[17] I am here using the concept of normal unconscious functioning in contrast to the "boiling cauldron" or "horse and rider" theory which regards unconscious processes mainly in terms of a raging source of irrational impulse, conflict, and psychopathology, which can be harnessed and made reasonable only by the ego and its defense systems.

ulatable fields of force. To speak about them is like speaking about an electromagnetic field. The unconscious is creatively inarticulate, and is better left so. Failure to recognize this fact has created unnecessary difficulties in analyses of that important area of the experience of temporality, and of spatiality, namely their unconscious experiential substratum.

Protomental unconscious processes, then, are concerned with as yet unverbalized human feeling and desire, with wishes, will, and intent, with libido, with loving and hating impulses, with primal feelings of envy and gratitude, of potency and omnipotence, of generosity and selfishness—with, in effect, everything that makes the world go round. They lack only the abilities connected with consciousness, to formulate things explicitly—albeit retrospectively—and to employ self-conscious reflexive logical and analytical skills. In effect, in the human being we have the spontaneous protomental unconscious process flowing about and constrained in a conscious logico-critical context.

Perhaps the most important thing about these protomental processes is that they are actively orientated toward something, toward doing something, changing something, achieving something, overcoming something, reducing tension, getting satisfaction, influencing others to feel something or to do something. They are, in short, always goal-directed, intentional.

It is this goal-directedness, this intentionality, of the unconscious protomental world which gives us our direct and immediate sense of temporal directionality. For we may now notice that one of the very significant features of a goal is that it not only states the will or intention to do something or to achieve some state or some condition, but it also includes the intention to do so by a certain time. Goals and intentions are organized in terms of time. They must be. Without a time by which an intention is to be carried out, a goal achieved, there is simply no goal or intention. To intend to do something some time or other, to achieve a goal but by no particular time, is the opposite of intention and will. It is to be will-less, intention-less, goal-less, floating, lackadaisical, spineless. The target time of completion, the willed time of achievement, the by-when it is intended to get somewhere, is as much a property of intention, of goal, of will, as is the redness of a red apple.

Because goals or intentions must *de facto* have target completion times in order to exist, we are led to the conclusion that unconscious protomental processes are impregnated with a powerful sense of time, saturated with it, dripping with it. Without quite being consciously aware of how it happened, we find ourselves urgently pressed toward reading the newspaper immediately, or with the intention of completing an essay by the end of the week, or with the plan to have a flat redecorated within the succeeding two months. We might change our targets as we proceed, but without a target completion time, even if it is continually changing, it is impossible to organize to get anything done at all.

We may, of course, not carry out the bidding of our unconsciously derived impulses and intentions. And we may modify those impulses and intentions as they become conscious and we are able to adopt a critical attitude toward them. But it is from the protomental reservoir that intentionality springs and in which our sense of the directionality of time thus has its source.

I would put this unconscious sense of temporal direction into words, that is, paraphrase it, in the following way: "I am here, now, at 7 P.M. on October 26. I feel slightly uncomfortable, and have the vague sense that I should be somewhere else. I remember that a week ago I received notice of a meeting which I intended to go to. I picture myself at the meeting and desire to go to it. It starts at 8 P.M., in an hour. I decide that if I get my car and drive to the meeting at once, it will be possible to be there on time. I will myself to go. I am ensconced in present desire, present memory, and present intention."

The effect of this unconscious experience is that at the time of decision, now, there is a past—present—future time axis which points right through my present experience at the moment. Its sensed direction is from my memory of the notice of a week ago of the meeting, through my desire to go, to my intention to be there in one hour. That is to say, this "arrow" of time flies from a week "in the past" to the present, to an hour "in the future." But its flight is nothing more than a mental fiction. It is an important fiction, however, representing as it does my active desiring, remembering, intending, willing, state of mind. Unconscious intention can be orientated from past toward the present into the future because both the past and the future are carried in the arms of the present. It could be represented in the accompanying diagram.

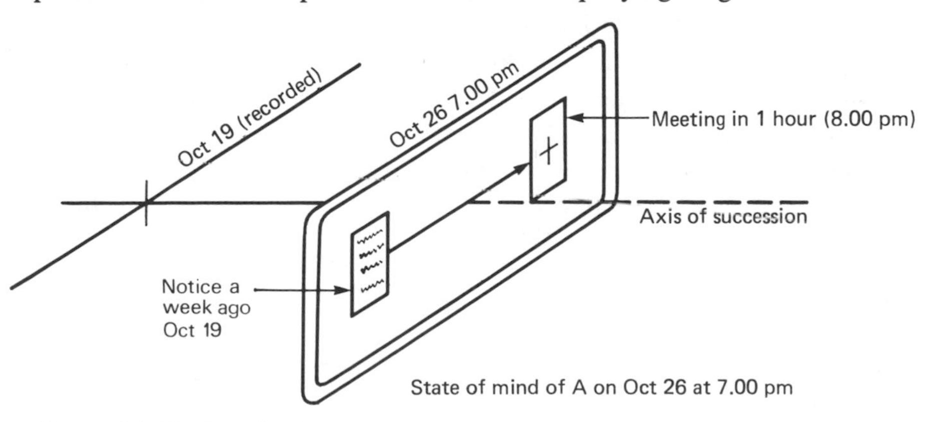

Diagram 4.1: Direction of an intent.

This unconscious sense of the dynamic from-the-past-to-the-future orientation of the field of force of will and intention in the case of specifically goal-directed behavior is reinforced by a more general sense of time connected with eternal beliefs and values. Such beliefs and values are felt as operating in a general way over long periods, or for life, or in the hereafter if that is a person's belief. They give a vaguer, weaker sense of temporal direction, however, than the powerfully sensed direction in time of goal-directed intentionality.

The Rationality of Protomental Unconscious Processes in Decision-Making

It may seem that my analysis suggests that as far as doing things is concerned, the protomental unconscious processes are the really substantial processes and the conscious processes are ancillary in the sense of being context-setting. That impression would be correct. It is important to recognize that it is the unconscious protomental processes which are fundamental, the conscious critical processes being an undoubtedly valuable adjunct. Verbal language, written language, explicit knowledge including science, conscious logic, computer competence, the cortical area of the brain, are all exceedingly useful to human beings, but they are not the central areas of life—or of time. The central areas are in felt rationality, felt goals, memories-in-feeling,[18] felt needs, lacks or absences which fill out and transform into wishes and desires and then overflow into action.

These felt or lived processes are not localizable in the brain or central nervous system. Nor are they to be localized in something called 'the mind' (which then has to be awkwardly placed somehow in 'the body' which is supposed to contain it). They are a function of the whole living organism with its sensory-motor system, its thinking and acting, and its touch-and-feel peripheral nervous system as well as its central one. For the protomental unconscious sensing processes in reasoning and decision-making are a matter of touch-and-feel, of green fingers, of skill, of nous, of good sense, of judgment, of discretion and discrimination, of shouldering uncertainty[19]—all good solid processes well insulated from too much tampering and interference from the conscious (although sometimes

[18] This is a concept vividly used by Melanie Klein in her psychoanalytic work with children. It describes the manner in which two- and three-year-olds respond to interpretations about past experiences in preverbal infancy, when they clearly get the sense of what is being said, in memory-feeling rather than in verbalized or pictorial memory. Adults also live continuously with such feelings, but they tend to be hidden from conscious awareness by the overlay of verbalized ideas which cover up so much of the richness of experience. See, for example, Melanie Klein, (1975), *Narrative of a Child Analysis* and *Envy and Gratitude*.

[19] See my analysis of uncertainty in Chapter VII of *Measurement of Responsibility*.

the insulation weakens and impoverishment of personality and of human contact is the consequence).

This view of protomental unconscious sensing processes as a central feature of reasoning and rationality is, of course, not a very common one.[20] I believe, however, that this is because of the widespread tendency to overvalue conscious mental processes, visual perception, and language. Thus, even Freud repeatedly said that the unconscious has no knowledge of time. For example, he stated: "The processes of the system Ucs. are timeless; i.e., they are not ordered temporally, are not altered by the passage of time, in fact bear no relation to time at all. The time-relation also is bound up with the work of the system Cs."[21] This view is one which has been elaborated by Marie Bonaparte in her article "Time and the Unconscious."[22] She adds, however, that "The statement that the unconscious is timeless may mean that the unconscious has no *knowledge* of time, that time as an intellectual concept does not exist for it. But to say this is to state a truism. The unconscious, the primitive reservoir of our instincts and our will to live, knows nothing of any concept; these are later acquisitions of the intellect. Consequently, it can no more have knowledge of the concept of time than of any other concept."

It may be seen, however, that both Freud and Bonaparte are reifying time. In the consciously formulated sense in which they are using the concept of time, it can of course be known only consciously. But then, of course, it is equally true that in this consciously formulated sense the unconscious has no knowledge of space either; or of triangles, or of circles, or of redness, or of any other verbally articulated concepts. The unconscious has only a nonformulated, nonconceptualized sense of past—present—future, of memory—perception—desire, and I see no reason to think that this organization is not present at least from birth. (This point is elaborated in my descriptions of the temporal frame of infant behavior in Chapter 9.)

[20] Polanyi developed the concept of tacit knowledge to deal with this problem. Thus, for example, he attempts to resolve by this concept, Plato's paradox set out in the *Meno,* namely, "to search for the solution to a problem is an absurdity; for either you know what you are looking for, and then there is no problem; or you do not know what you are looking for, and then you cannot expect to find anything. . . . The kind of tacit knowledge that solves the paradox of the *Meno* consists in the intimation of something hidden, which we may yet discover." M. Polanyi (1966), *The Tacit Dimension,* pp. 22, 23. I believe that Polanyi would have found himself on stronger ground with the conception of unconscious mental functioning.

[21] *Standard Edition,* Vol. IV, p. 185.

[22] Marie Bonaparte, "Time and the Unconscious," p. 439.

This notion of the rationality of unconscious mental processes will be of substantial importance for my analysis in later chapters. In particular I shall apply it to the time-span of decision-making behavior, and demonstrate the significance of the understanding of the unconscious time sense for the understanding and assessment of human competence.

Decisions as Time-Targeted Protomental Unconscious Processes

The rationality and field-of-force organization of the unconscious memory–perception–intention temporally directed field shows clearly in decision-making. Decisions are always goal-directed, since a choice can be made only in relation to an intended outcome. Moreover, deciding or choosing are always in the final analysis founded upon unconscious processes. The role of protomental unconscious factors in decision-making can be succinctly summed up in the fact that if you consciously know all the considerations that led to a particular choice then you did not make a decision: you were a computer. Let me explain this statement.

The making of choices and decisions is connected with *uncertainty*. If, for some reason, you can be *certain* that a particular goal is called for, or that a particular path to a goal is the optimum, then there is no decision to be made: unless, of course, you wish to decide—for some reason or other about which you would assuredly not be certain—not to seek the required goal or not to take the optimum path toward it.

The factors and possibilities entering into human choice situations are always infinite. The reasons are infinite, regardless of whether the options are finitely limited, as in, for example, gambling on the toss of a coin, or infinitely open, as in most choice situations with respect to action in real life in solving a problem or in satisfying a need. There are always the choices that have been made previously in similar (but never identical) situations; there are all the reasons which have never been thought up before; then there is the possibility of postponement; or of giving up and trying something else; and finally there are all the conflicting desires, aspirations, expectations, hopes, fears, prejudices, influences from others—some conscious and some unconscious. It is out of this amalgam of the formulated, the unformulated, and the unconscious that we somehow decide what to do, sometimes quickly and sometimes with hesitation, sometimes with a confident feeling of a high probability of a satisfactory outcome and sometimes with a pessimistic feeling of low probability. It is because of this inevitable uncertainty and play of unconscious factors that some modern decision theory operates on the basis of establishing probability judgments and of helping decision-makers to formulate their view of the degree of uncertainty of their judgment in terms of their sensed probability of a particular outcome.[23]

A judgment of probability is a rational and not a random process; it is an unconscious process made within a conscious context. The context and the outcome of the unconscious judging process usually become conscious, but not the process itself. Conscious knowledge can help to limit the field of choice within which we apply our unconscious rationality. It provides context. It also provides a logico-critical context for the evaluation of progress toward solutions as well as of final solutions, helping to eliminate errors. But it cannot make decisions.

To bring certainty into decision-making is a contradiction in terms. It would be necessary to bring all the conditions under control, to confine the situation rigidly and narrowly to a specified number of specified alternatives at each choice point, and to program in explicit terms what is to be done for each choice. Such a procedure has the inevitable effect of removing all the elements of decision. It is this type of control which characterizes the computer.[24] What cannot, however, be built into a computer is what constitutes the decision process; namely, the nonformulated, nonexplicit, nonexplicable, unknown, nonverbalized play of unconscious forces: that is a property of living organisms only.

In short, then, we have the unconscious sensing of the temporally directed past—present—future field in the unconscious interplay of memory—perception—desire—intention. We have the awareness of motion, of process, in preconscious background mental activity. And we have the atomistic points of objective time and earlier and later occurrences in the focus of conscious and verbally formulated mental activity. In the interplay of the three processes—conscious, preconscious, and unconscious—we get the human being living his life in space and in time, oscillating between states of contemplation and action, with his knowledge of what came first and what came later, his more limited knowledge of his motives and goals, his reliance upon his vaguer awareness of what is going on round about him, and his dependence upon his unconscious protomental sense of striving and intentions. I shall turn now to consider further the manner in which he organizes his dynamic past—present—future field at successive points in objective time.

[23] See, for example, L. D. Phillips, (1974), *Bayesian Statistics for Social Scientists;* P. C. Humphreys and A. Wishuda, (1979), "Multi-Attribute Utility Decomposition"; L. D. Phillips, (1980), "Generation Theory."

[24] The programming of random elements into a computer in no way changes this argument. Human behavior, human choices, human decision-making are never random even in so-called trial-and-error learning. Whatever is going on, there is always all-pervasive purpose suffusing the psychological field, as gravity suffuses the physical worlds. When computers can love, get bored, feel bereaved, become over-eager and impatient, have their decisions influenced by their own personal sense of responsibility and conscience, suffer disappointment, and experience success and triumph, then they too will make decisions. But they will have become human beings in the process.

CHAPTER FIVE

The Perpetual Present

The idea of two different types of time crops up continually in our investigation. These two types are the time of succession—the atomistic, point-at-a-distance, discontinuous, clock time of earlier and later, of the physical world; and the time of intention—the unconscious, protomental, continuous memory—perception—anticipation field-of-force time of past—present—future. I shall argue that our conception of the human world requires a two-dimensional temporal construction, just as our conception of the physical world requires only a one-dimensional temporal construction. The two temporal coordinates are those of succession and intention.

Before considering this two-dimensional temporal world, however, it will be necessary to complete my analysis of the present, both in its own right and in its containment of past and future. In doing so, I shall discriminate three meanings of the present: the present of the immediate flowing now; the present of the things in which I am actually engaged; and the present of my life as a whole, of my existence; or, in short, the immediate present; the active present; and the existential present. From this stand it will then be possible to examine the processes of retrodiction and prediction as probability statements of what "the future might really be like."

Out of these considerations we may then establish our two temporal coordinates. That will in turn put us in a position to develop the use of a workable theory for the formulation of intentions and goal-directed or purposive behavior, for the exploration of the social and psychological worlds that individual human beings live in (Karl Popper's World Two and World Three), and for the assessment of the size of that social world and of the capability of the individual. These themes, along with the application of our analysis to the statement of the nature of social and psychological phenomena, will be pursued in succeeding chapters.

The Immediate, the Active, and the Existential Present

I have so far been using the concept of the present that refers to the immediate or the flowing present. This immediate present has been described in many different ways by many authors. It is the "present bit" of time, or "specious present," or "sensible present," or "saddle-back present" of William James,[1] or the "moving present" of Stearns,[2] or the "actually present" of Koffka,[3] or the "mental present" of Pieron,[4] or the "perceived present" of Fraisse,[5] or the "passing present" of Dennes,[6] or the "travelling now" of Santayana,[7] or the "moment that is creation and fate" of Tillich.[8]

But this very immediate flowing presentness (which I shall examine in detail in the following section) is not the only meaning we give to the present; life presents itself in other ways as well. The concept of immediate present refers to one region of a field of experience which comprises the continuously interpenetrating regions of past—present—future in simultaneous memory, perception, and intent. But it is possible to differentiate and to abstract and refine a whole range of other things called present, as, for example, the present moment, the present day, the present year, the present decade, etc. It is like dividing a country into the northern region, the central region, and the southern region. Where the boundaries are drawn is an arbitrary matter of convenience, and the boundaries can be changed—or dropped—*ad lib*. For my purposes I shall find it useful to categorize two other regions of the present, in addition to the immediate present. These two are: the active present and the existential present.

By the active present I shall refer to the extended present in which each person lives his active life. It is the presentness of the things which we are actively doing. It will vary, will be different, for each one of us. It is what a person began yesterday and hopes to finish next week, or began last month and hopes to finish next month, or in the case of longer-term activities engaged in by some people, might have started last year with a difficult and complex program planned to be completed in a year or even in a few years' time.

This active present also has the feeling of "nowness," even though it extends over a longer period of time than the nowness of the immediate

[1] William James (1890), *Principles of Psychology,* Vol. I, pp. 608 and 609.
[2] I. Stearns, (1950), in *Review of Metaphysics* 5, p. 198.
[3] Kurt Koffka, (1935), *Principles of Gestalt Psychology,* p. 437.
[4] H. Pieron, (1923), "Les problémes psychophysiologiques de la perception du temps," p. 9.
[5] P. Fraisse, (1964), *The Psychology of Time,* p. 85.
[6] W. R. Dennes, (1935), *The Problem of Time,* p. 103.
[7] G. Santayana, (1934), *Realms of Being.*
[8] P. Tillich, (1936), *The Interpretation of History,* p. 254.

present. It is what Stuart Hampshire has described as the continuous present. He writes: "The stretch of measurable time that may be referred to by the word 'now' is as indeterminate and elastic as is the range of 'present action' or 'present situation.' There could be no temptation in experience to think of the present as a point instant. The 'continuous present' is the name naturally given to the tense that we use in describing our actions contemporaneously."[9]

The active present defines or circumscribes the working individual: working in the sense of pursuing a specified goal—whether doing something physically, or engaged in mentally working out a plan or the solution to a problem. It is to be contrasted with the timeless, presentless sensation which accompanies aimless musing or open-ended contemplation or reflection. I shall consider this active or working present in Chapters 7 and 8 in establishing the concept of the purposeful or goal-directed episode, and the concept of the temporal frame of the individual. I shall there develop the theme of how each individual constructs his own personal active present in an idiosyncratic way, how he sets the boundaries of this active present for himself, and how the boundaries are sometimes set for him by others.

Beyond the active present of the person, the present in which each one carries out his work, there is our sense of the presentness of our life as a whole. This presentness is contained in our sense of intactness and continuity as a person. It is the sense of being the same person one was as a child, and of continuing to be the same person as one goes on changing, as one grows older.

This lifelong presentness is the presentness of coexisting Being and Becoming, the present of existence, the existential present. It is the present of the self, of the person, the present which is the timeliness of one's personal identity. The existential present is the living outer boundary which frames our immediate present and our active present, giving to both of them their personal meaning, their sense of being parts of our own lives.

Still further beyond the existential present of human being there is a time which I would hesitate to term "present." It is the time of history and of eternity, extending limitlessly, endlessly, into the vast reaches of worlds beyond our ken. At their outer boundaries these worlds become mysteries which are present only in the sense that each individual gives or finds his own personal metaphysical meaning for them, or gives them no meaning at all. Let me then draw back from these obscure and mys-

[9] Stuart Hampshire, (1959), *Thought and Action,* p. 72.

terious regions, and return to the more substantial and seemingly tangible world of the immediate now.

Some Characteristics of the Present

We have seen that it is seriously misleading to talk of future events as though they existed as external things other than as probability assumptions in the minds of individuals about things which might or might not happen. There is no such external thing. There is no such thing as an external event which somehow exists now in the future, which is traveling toward the present and which will be present at a future time, after which it will move out of the present toward the past and become a past event.

It is a play on words to say (retrospectively) in 1950 that for the people who lived in 1930 World War II "existed" in the future. It did not, at least not in the sense that World War II then existed and was somehow flowing or moving toward them. In 1930 there was 1930: there was no World War II, in the past, in the present, or in the future. By 1950 what then existed was 1950, plus records indicating a succession of events dateable at such years as 1930 and 1939. The only correct statement in 1950 would be that World War II started nine years later than 1930. Moreover, in 1950 it would be meaningless to say that 1960 would bring certain events already fixed but unknown in 1950. This form of thinking is crude mechanistic materialist determinism fueled by the magic of self-fulfilling prophecies.

The flaw in the future-to-present-to-past treadmill concept of time as flow or passage (and note that we are back to McTaggart's unfortunate A-series formulation) is that it takes the concept of time as succession of events, reifies it, transforms it by magical anthropomorphism into a human guise, and confuses that conception of time with human desires and memories expressed in time terms. These latter are the part of the story that St. Augustine nearly got right—except that he identified memory, perception, and expectation with time rather than seeing them in relation to time or in terms of how they give rise to the idea of time. Past, present, and future are therefore not different kinds of time either separately or together. They are modes of organization of our current mental experience in terms of time.[10] In order to understand them more sharply, let us examine each in turn.

[10] Martin Heidegger in *Being and Time* analyzes time as a field in which past, present, and future are held together as a unified system. He lays great emphasis upon future as the synthesizing element in giving meaning to the past—present—future field in terms of human intention and desire. But his analysis is couched too much in conscious terms, so that the sense of temporal field tends to disintegrate, while at the same time there is no room for an atavistic objective formulation of time. Martin Heidegger, (1962), *Being and Time*. See also my reference to his analysis of dimensions of time in his later work, *On Time and Being*.

Consider first the concept of past things. It has two different and important meanings. First, in the internal world of memory it means "I remember something and what it meant to me personally"; second, in the external shareable world it means "There is this record (documents or artefacts of some kind) that an event A happened earlier than now or earlier than B which is earlier than now."[11] That is to say, there is the pastness of internal memory and meaning; and there is the pastness constituted by a dating from extant records—the past of historical reconstruction and of interpretation of the findings.

If we then consider the concept of things future, we again come upon the two meanings: first, that of human intention or prediction, "I *intend* to buy something"; or "I predict from these artefacts that A will happen"; and second, that of recorded earlier and later succession, "The records indicate that when A occurred B had still to occur two days later."

The pastness or the futureness of two events can be abstracted, made conscious and static, and verbalized because they are either retrospective or prospective, and can retain their fixed relationship with each other of so much earlier than or later than. It should be noted, however, that they cannot have a fixed relationship of so much earlier than or later than now, because that relationship is continuously changing—it is totally impermanent, hence totally unformulatable. Earlier than and later than are thus the objective modes of dating the relationship between any two temporal points. It is for this reason that tensed language can be reformulated with respect only to objective events, and can give a fixed and permanent relationship: that is to say, "A happened three years before B" becomes "There is an event A and it happens on a day three years earlier than B happens"; or when we way "Two days ago" we in fact mean "forty-eight hours earlier than just one moment ago when I had the experience of two days ago."

When we turn to the concept of things present, however, the picture changes, for only one meaning is available and that is the sense of an internal and unconscious meaning. There can be no external and objective meaning attributable to the idea of the immediate now, the immediate present. There is no record or artefact to refer to with respect to pres-

[11] It is for this reason that tensed language can be eliminated in formal logic as in Quine. It can be eliminated because formal logic is the expression par excellence of a consciously formulated activity. As such it is expressible solely along the temporal axis of succession, in terms of time of earlier and later. What it cannot do is deal rigorously with human intention, including the intentions of the logician himself at work. Modal logic attempts to deal with this problem and does so to some extent, but it cannot deal with the fluctuation, fluttering, inaccessibility of the flowing unconscious experience that is the actuality of what uncertainty and probability are about. Willard van Orman Quine, (1960), *Word and Object*.

entness; no comparison between the immediately present A (now) and a past B, nor between the immediately present A and a future B. For, the moment we have abstracted and verbalized and made conscious a particular A, the thing or event described is earlier than the formulated description. In short, the immediate present cannot be dated in the immediate present, because as soon as it is, it is no longer in the present. That is the point emphasized by Bergson.[12]

In contrast, it is possible to be aware preconsciously of the present and to sense it unconsciously—but, note, without verbal formulation either subjective or objective. It is this awareness and this sensing which give the ongoing flowing unbroken background both to Being and to Becoming. So long as we leave it as the background to focused knowing, we can have the sense of presentness. Bring this sense into conscious focus and make it verbalizable, and all we can do is report, "At a moment just earlier than this moment now when I am speaking about it, I remember that I was aware of my Being, aware of the present, and I unconsciously sensed the present." A person's preconscious awareness and unconscious sensing is always an awareness and a sensing of the ongoing state of affairs in that person's mind "now."

There need be nothing mysterious about this apparent disjunction between the external objective dichotomy of time into earlier and later, and the internal unconscious experience of past, present, and future as one unified field of force, and their external objective conscious experience. We have seen that the objective mode of experience is an atomistic, verbalizable, and static mode of abstraction. It cannot, therefore, refer to "that present" because the immediately experienced present cannot be culled, pulled out, abstracted, formulated, and verbalized in its own present but only at a later point, in retrospection. The immediate "now" is an unconsciously experienced flow of events—the flow experienced as life itself.

When we consider the direct experience of the present, therefore, we return to the inexpressible, the nonverbalizable. It is embedded in the total unconscious field of immediate onflowing experience of past—present—future, memory—perception—intention. The immediate present is a description of the characteristics of a mental field. It is not time which is flowing; it is not future which is flowing into present, into past. It is one's psychological state, one's mental activity, which gives the feeling of flowing. But we can now supplement my statement in the previous chapter that our sense of flux is a preconscious awareness of motion. Flow, flux, is grounded in the unconscious experience of the functioning

[12] Henri Bergson, op. cit.

of one's own mind, including the unconscious experience of the continuous unfolding, shifting, and changing of goals, desires, memories, intentions, and feelings.[13] There is passage or succession in events, in processes, in the external world, and in the internal mental world, but no passage of time. We ourselves are living processes; we ourselves, therefore, are in passage—from birth to death, from a datable day of birth to an as-yet-to-be-dated end. It is because death is contained in time, unavoidably bumped into as part of our intentional and existential time future, that time is emotionally more upsetting than space.[14]

It is because the immediate experience of the present can be only an unconscious or a preconscious mental experience that there can be no single answer to the question of how long is the immediate present, of when it begins and ends. For it does not begin or end. It is one part of a continuous unconscious or preconscious field, and as such has no delimitable boundaries. To try to get a single answer, therefore, to the definition of the limits of the immediate present is like trying to get an answer to the question, "How high is a piece of hate?" or "How many hours is a piece of ribbon?" Just because we cannot answer such inappropriate abstractions does not mean that hate and ribbon do not exist. It merely means that we have asked the wrong question. "How long is the immediate present?" is a wrong question. So are "How long is the past?" and "How long is the future?" If we stopped asking such questions we could avoid such other questions as: "If there was a beginning of time, what was going on before then (or after the ending)?"; and "If there was no beginning of time, how could that be?" Like the paradox about the village barber or Zeno's arrows, such questions may perhaps be useful as puzzles with which to advance our knowledge of logic and of other mental processes; but they are certainly not answerable as questions about the world around us.

Immediate presentness, then, is the present of immediate sense and awareness. It is a dynamic present of where we are as we live and breathe, not from minute to minute nor from second to second, nor from micro-

[13] "I could not even begin to conceive what it would be like not to have this immediate sense of before and after in movement and action, as the condition of all my experience as agent and observer. Because I always have intentions, and because knowing what I am doing at this moment necessarily involves knowing what I have just done and knowing what I am immediately about to do, my attention does not rest in the present. . . . Ordinarily we carry our intentions with us, and this carrying forward of intentions, together with the perception of movement, provides the natural and necessary continuity of experience. Any action, as an intended bringing about of an effect, has a certain trajectory, a relation of before and after within it." Stuart Hampshire, *Thought and Action,* p. 72.

[14] I have described how this personal awareness of death takes hold of people in midlife. For me it is the main explanation of the occurrence of the intense emotional reaction in midlife which I have called the "mid-life crisis." E. Jaques, (1965), "Death and the Mid-Life Crisis."

second to micro-second, nor yet even from the most infinitesimally short (a contradiction in terms, I know) moment to moment, for it is the flowing unbroken continuous present.[15] It is in this flowing quality that it differs from the active present, which is an abstraction, a discontinuous slice of time describing an episode with a selectively defined beginning and a selectively defined end. It is this immediate present which is the present of the unconscious mental process which is the seat of our uncertainties, of our judgment, of our intentions, of our decisions, and—as I shall now describe—of our interpretations of our sense of past and future as these appear in prediction and in retrodiction.

Prediction and Retrodiction

Given that past, immediate present, and future are all part of the current psychic experience of the individual, then retrodiction is not necessarily all that much more certain than is prediction. Stuart Hampshire has expressed this idea very cogently: "It is sometimes argued that the past is that which is necessarily fixed and unalterable, and which can therefore be the object of certain knowledge, while the future is that which is necessarily unfixed and alterable, and which therefore can never be the object of certain knowledge. . . . If the objection is pressed—'But something might happen at any moment in the future which might make me change my mind,' the same can be said about the past; something might happen at any moment (perhaps new testimony from others) to change my mind about the past."[16]

Certainly, to take one example, the Darwinian revolution was a most massive shifting of our interpretation and understanding of the origin and historical development of species. Such a conception of our historical background is based upon the observation of things and events, and of artefacts and records, in the current world. This perception of our current world is then *interpreted*. A part of this interpretation is the retrodiction of certain events, and an earlier dating of those events. While these retrodictive interpretations may be interesting descriptions of the kind of thing that might have taken place, or that probably took place at a given earlier date, they are not to be confused with a factual picture of what really went on in the earlier event of a so-called real past.

Accurate interpretive retrodiction is possible only when events occur under highly controlled conditions and are recorded as completely as possible. The more complete the control and the more accurate the at-

[15] As Guyau points out: "We must desire, we must want, we must stretch out our hands and walk to create the future. *The future* is not *what is coming to us* but *what we are going to.*" J. M. Guyau, (1890), *La Genèse de l'Idée de Temps:* p. 33.

[16] Ibid., p. 127.

the-time recording, the more accurate the retrodiction can become. But so also does prediction become more accurate, and probability judgments become higher, with higher degrees of control. Neither perfect records nor perfect control is possible: therefore neither one hundred percent prediction nor one hundred percent sure retrodiction is possible.

Even Aristotle's sea battle does not necessarily have to take place tomorrow or not take place tomorrow. There is a high probability that things will be either A or not-A, but never an absolute certainty. It all depends on how the human mind will be working tomorrow. For a prediction is always a verbal statement of an imaginative idea. The actual event which succeeds a prediction at a later date is not a verbal statement: it is a perceived whole event, directly experienced, which then has to be verbally formulated. The verbal formulation both of the prediction and of the actual event are grossly incomplete expressions of the ideas and perceptions they are supposed to map. How such formulations then match each other—the recording of the verbalized prediction versus the verbalized experience—must always be problematic.

It is not possible, for example, to know what will happen to the concept of sea battle or of any A about which a proposition might be stated. This kind of possible ambiguity is precisely what much of the law is about: was what happened today the same kind of thing that would have been called a sea battle yesterday, or was it not? Was what was done a crime or was it not a crime? Did a man run across a neighbor's lawn in order to give chase to a thief in such a way as to constitute a trespass or was it not a trespass? There can never be certainty, because there can never be certainty about how the words today to state a prediction or to record the interpretation or memories of a previous event, will be understood when checking the prediction if and when tomorrow comes, or were understood at some time in history as compared with now.[17]

Association of Need, Memory, and Intention

The starting point of a description of temporal space is the immediate or perpetual present—for it is around this perpetual present that the temporal space is organized.[18] The perpetual present contains, in the first place,

[17] Bergson, of course, trenchantly argues this case as part of his conception of "the continuous creation of unforeseeable novelty which seems to be going on in the Universe." As he describes it: "No matter how I try to imagine in detail what is going to happen to me, still how inadequate, how abstract and stilted is the thing I have imagined in comparison to what actually happens! The realization brings with it an unforeseeable nothing which changes everything." H. Bergson, (1965), *The Creative Mind,* p. 91.

[18] William James describe this immediate interplay of memory and anticipation around present perception as the "ownership" or "appropriation" of past and future by present thought. He sought by this formulation to get away from Association theory and from Hume's notion of past, present, and future as discontinuous percepts following one another in time and connected by the cause and effect of association. William James, (1890), *Principles of Psychology.*

that most significant aspect of all action, namely the experienced sense of lack, the constructed emptiness, the nothing, the gap, the sense of something missing in life, which is experienced as need, desire, the wish which seeks fulfillment and arouses fantasies and action.[19]

On the other side of the inner world—external world divide in the perceptual present is the external world perceived as a storehouse of utility—filled with people and things any or all of which might be of value in filling the empty—the gaps and lacks—whether these gaps and lacks be for things to be obtained for oneself or for things to be done for others.

The sleep of memory is disturbed by need, and dormant ideas and experiences surge forward to press into the present. What memories will burst into the foreground is unpredictable. Some will be pulled forward by an unconscious matching with current need—a pattern or configuration of memories which contain possible means of overcoming present problems. Other memories of unresolved experiences will push themselves forward into the present arena, seeking outlet and resolution in and through attachment to the current path of action.

This gestalt or configuration of memory and perceived need—an unformulated principle of *prägnanz* is at work—reflects the true functioning of mental association. It is not an association in which there is a passage between "past" and "present." It is a construction arising out of particular associative patterns or amalgams of present perceptions and the field of present memories and desires and intentions. The interaction between association bringing forward remembered knowledge and experience to help with current problems and to guide action, and association opening up the release of memory systems under the tension of frustration, is shown in the phenomena of transference in psychoanalysis, and in the Ovsiankina-Zeigarnik effect in problem-solving.[20]

Similarly the field of anticipation, of expectation, of seeking, is linked to the filling of a current gap. Under the impact of the absent, the nothing, the missing, there is a mental construction of the something-yet-to-be, which will fill the gap and overcome the empty and the nothing.

[19] This emptiness or sense of gap which is the source of need and action is at the root of Sartre's concept of négatité or nothingness. "Négatité est été" (Nothingness is made-to-be) he writes in *Being and Nothingness* (1956). Or, as Danto paraphrases him: "Pierre's not being there [Pierre and the café] is my responsibility and not Pierre's." Danto, (1976) *Sartre*, p. 68. And again, "Nothingness is a kind of shadow which we cast rather than a vacuity which we discover." Ibid. p. 70.

[20] See M. Ovsiankina, (1928), "Die Wiederaufnahme von unterbrochenen Handlungen"; and B. Zeigarnik, (1927), ":Über das Behalten von unerledigten Handlungen." When subjects in problem-solving situations returned to the situation in which they had been confronted by the problems, they tended to remember and to return spontaneously to those problems which they had previously failed to solve, rather than to those they had resolved.

This something-to-be is the goal. It is the mental model for the thing which will possibly give a mental state of gratification and resolution if and when it is realized in fact.

The goal is thus future only in the sense of being a mental representation of something which might give satisfaction if it were to become available. It is a state of mind—this state of future exists in the inner world of the present. The satisfaction will be brought about, however, by the creation or achievement of something in the external or the internal world; that is to say, by the transformation of something in the present storehouse of utility into something which will fulfill the need. The presently future goal thus is made up of the state of affairs intended to be created, and of the state of mind which it is envisaged will ensue from the new state of affairs.

Two things may now be further clarified. First, the past, present, and future are simply the conscious concepts we construct to express our conscious sense of our unconscious experience of oscillation between a focus upon our intentions, upon current memories, and upon current wishes or desires. There can be no clear-cut boundary between them, because they can never be present in focus simultaneously with a boundary between. They are subregions within one current unconsciously—and to some extent preconsciously—experienced field at one and the same point of succession. When one or other subregion is in focus, in figure, the other subregions constitute part of the preconscious ground.

Second, the field of memory—perception—anticipation is continuously changing. There are no fixed memories, no fixed perceptions, no fixed goals. Memory, perceptions, and goals are continuously changing and reorganizing. They are processes.

This flux is best illustrated by goals. A particular goal may seem to be relatively fixed and constant as we work toward it. But a moment's reflection will show that, in fact, a goal must be continuously undergoing change as we work toward it: it becomes increasingly well defined; it is modified; it becomes more or less attractive, or accessible, or desirable; it may be dropped. It is to this question of goals, purposeful behavior, episodes, that we shall now turn our attention.

PART THREE

DIMENSIONS OF THE LIVING WORLD

CHAPTER SIX

The Two Dimensions of Time: Succession and Intent

We now reach the central theme of my argument, the fulcrum, the formulation around which the whole revolves. It is concerned with the question of how human beings organize their temporal world. Having put aside the notion that time itself has directionality, that it moves or passes from future to present to past, I shall replace such notions with a two-dimensional formulation of time and a five-dimensional world of human action.

These two dimensions of time are: the dimension of succession—which contains the conception of a historical reconstruction of temporal points in earlier and later relationship to each other; and the dimension of prediction and intent—which contains the conceptions of goal-directedness and of what will happen, in the continuously present field of past—present—future which coexist in the interaction of memory, perception, desire, and anticipation.

Only one of these two temporal dimensions—the dimension of succession—is needed for the construction of the four-dimensional world, even the physical world of relativity theory. But, as I shall show in later chapters, the two temporal dimensions are necessary for constructing an adequate theory of the social and psychological world, complete with purpose and intention and human agency, including the natural scientist's predictions about the physical world.

Although I shall be concentrating upon the problem of the dimensions of time, this focus upon time does not mean that space does not enter into the location of desires, predictions, intentions, and goals. Even though intentions do not themselves have length, width, and depth, nevertheless intentional behavior must be localizable in space as in time. It must be conceived of as occurring at some specific place as well as by some specific time, otherwise it cannot exist. To intend to have lunch, or to carry out an interview, or to become a lawyer, or to get a job, is to plan and seek to eat, or interview, or study, or work, not just anywhere but in some specific place or in some few possible places; unless, that is, there is no real intention but only a vague desire to do something, somewhere, some time, but who knows what, or where, or when?

The world of human action, prediction, intention, purpose, meaning, is therefore a five-dimensional world. Or perhaps, as Herman Weyl and J.R. Lucas[1] might suggest, it is best expressed as a (3 + 2)-dimensional world, keeping separate the three Cartesian coordinates locating the place and the two temporal coordinates locating the planned time and achieved times as particular times or dates. It is in this five-dimensional world that we live as human beings.

Before constructing the two-dimensional temporal theory, I shall first consider the general nature of dimensions by examining the dimensions of space (spatial coordinates). Then I shall examine the four-dimensional temporal theory of Heidegger and the unidimensional temporal theory of Williams, to provide a framework or setting and to provide a background of comparison.

Dimensions and Coordinates of Space and Time

I shall use the concept of dimension in the same manner as it is used in organizing our ideas of space. It carries the meaning of the coordinates along which we measure the extension and position of points and objects, thereby identifying and locating them. The fact that space is space—a univocal concept—does not mean that we cannot orientate toward it in many ways. In primitive behavioral terms we cognitively organize our outlook, we locate ourselves, we move about, in a three-dimensional world of up, down, and sideways, conceptualized in the Cartesian 3-coordinate system.

If we then ask why we approach the spatial world as we do, the answer must be that it is because we are built that way; because we have constitutionally given gestalten of length, breadth, and depth, by means of which we organize the spatial cognitive field. And we are also built in such a way that we can function conceptually, as Euclid did, in a two-dimensional theoretical world of plane geometry,[2] or in any number of dimensions from the zero-dimensional to the n-dimensional worlds of

[1] Weyl and Lucas prefer to speak of a (3 + 1)-dimensional world rather than a four-dimensional world of relativity. Lucas, for example, writes: "The distinction between space-like and time-like dimensions remains, and is essential even in relativity theory, and is best indicated by our speaking of (3 + 1) dimensions." J.R. Lucas, (1973), *A Treatise on Time and Space,* p. 194fn.

[2] The world from which the term dimension—two-fold measure—was derived.

mathematics. How many dimensions we choose to employ will depend upon the problem, and may vary with our mode of analysis.

I shall attempt to establish that this same form of thinking is apposite to the consideration of time, that it is essential for its systematic understanding. The construction of temporal coordinates is as necessary for locating events in the physical world and for locating human relationships and processes in the social world, as are spatial coordinates for locating things in a simple spatial model of the world. It is important to note, for example, that relativity theory has required the introduction of one temporal coordinate to combine algebraically with the three spatial coordinates for locating physical events; we shall need to be clear about how many coordinates may be needed for human events.

But how can we know how many temporal dimensions are in fact required? Martin Heidegger has argued for four; Donald Williams argues the case for one only. Let me cast their arguments for the moment as our Scylla and Charybdis respectively, both seductively interesting but each with its own shortcomings and dangers which it will serve our present purpose to understand.[3]

Heidegger's Four Temporal Dimensions

It was in his later work that Heidegger rejected the commonly used unidimensional physical world conception of clock time in favor of his own four-dimensional construction. The one-dimensional construct which he argued had nothing at all to do with time, was that of "the sense of the distance measured between two time-points . . . the result of time calculation. In this calculation, time represented as a line and parameter and thus one-dimensional is measured out in terms of numbers." [4] This

[3] Most authors tend to see time as either unitary or as a twofold notion. Those who favor the twofold notion are many, but they see the issue in terms of different approaches to time, or of different types of time, rather than as true dimensions or coordinates. Lucas, for example (op. cit., pp. 264–65), argues that "the modal approach to time is much more consonant with our view of ourselves as responsible agents than is the tenuous time of classical physics. It has an inherent direction." Then, in a most interesting vein, he writes, " . . . no one approach will enable us to comprehend the whole of the concept of time. The argument from consciousness, like the argument from change, secures the topological properties of denseness, continuity, and linearity; the argument from agency is less convincing on these scores, but decisive on the equally important ones of the direction and the modality of time."

Lucas here moves close to a two-dimensional construction. Moreover, he relates his second approach to agency and modality rather than to consciousness. Does he mean to link them, therefore, to the unconscious? It would appear not, for on page 310 he reverses his argument slightly, and links "personal [rather than physical] time as the concomitant of consciousness and as the condition of choice, and public time as the dimension of change." If he had made the linkage between time of agency and the unconscious, then I think he might have avoided the unnecessary allusion to time as passage, in his case in the "characterisation of time as passage from aspiration to achievement."

[4] Martin Heidegger, (1972), *On Time and Being,* p. 14

dimension is identical with the concept of the axis of temporal succession. But instead of keeping it as one of the dimensions, Heidegger argues that it is too spatialized to have anything to do with time at all and throws it away, an argument which would eliminate all possibility of locating either physical events or human events in clock time.

The four dimensions which he then constructs as necessary for understanding what he calls "true time" are the three dimensions of past, present, and future,[5] complete with passage from future to past, plus a fourth which resides in what he calls the unity, or interplay, of the first three dimensions: "The unity of time's three dimensions consists in the interplay of each towards each. This interplay proves to be the true extending, playing in the very heart of time, the fourth dimension so to speak—not only so to speak, but in the true nature of the matter. True time is four-dimensional. dimension which we call the fourth in our count is, in the nature of the matter, the first, that is, the giving that determines all." [6]

This analysis of Heidegger's is of interest for my analysis, in that his fourth dimension "which is the first" is precisely the one which I have argued is most readily comprehensible in terms of the concept of the unconscious field of past, present, and future, functioning as a true mental field of force. But Heidegger lacked a conception of unconscious mental processes, and was thrown back upon the notion of an ill-formulated "It" which gives time and space, the "It" which gives Being.[7] Whereas Freud's concept of the unconscious opens up the whole world of mental functioning, Heidegger's conception of a phenomenal "It" has a closing-down effect. For his fourth dimension is really not the fourth. It really encompasses past, present, and future, binding them into a single whole: not four dimensions, but one single dimension composed of three interacting regions. But Heidegger did not see the matter that way. In rejecting the dimension of succession and keeping past, present, and

[5] Past, present, and future tend to be regarded as three dimensions of time by phenomenology and existentialism in particular. Thus, Sartre, for example, stimulated by Heidegger, employs this conception in his chapters on temporality in *Being and Nothingness*.

[6] Heidegger, op. cit., p. 15. This quotation continues as follows: "But the dimension which we call the fourth in our count is, in the nature of the matter, the first, that is, the giving that determines all. In future, in past, in the present, that giving brings about to teach its own presencing, holds them apart thus opened and so holds them toward one another in the nearness by which the three dimensions remain near one another. For this reason we call the first, original, literally incipient extending in which the unity of true time consists 'nearing nearness,' 'nearhood' (*nahheit*), an early word still used by Kant. But it brings future, past and present near to one another by distancing them. For it keeps what has been open by denying its advent as present. This nearing of nearness keeps open the approach coming from the future by withholding the present in the approach. Nearing nearness has the character of denial and withholding it. It unifies in advance the ways in which what has-been, what is about to be, and the present reach out toward each other."

[7] Ibid., p. 17

future as separate dimensions, he ended up having lost clock time and trapped in the idea of a passage or flow of what was left of time from future to present to past.

Williams's Single Temporal Dimension

Donald Williams, on the other hand, argues trenchantly and effectively against the concept of time's passage—as against "real succession, that rivers flow, and winds blow, that things burn and burst, that men strive and guess and die. All this is the concrete stuff of the [space–time] manifold, the reality of serial happening, one event after another . . . ."[8] As we have seen, however, whereas serial happenings, or succession, can give us a location of objective events in clock time by dating them as earlier and later, they cannot really give us, as Professor Williams seems to believe they can, any satisfactory grip upon men's strivings and guessings, upon men's intentions. For this latter purpose it is necessary to bring in the past, present, and future, not, as Williams rightly argues, as passage or flow, but, as he fails to see, as coexisting in the present field.

Professor Williams himself makes an intriguingly similar analysis to our own of a second temporal dimension at right angles to that of succession, but only to throw it overboard. He vividly opines that, "It is conceivable then, though perhaps physically impossible, that one four-dimensional part of the manifold of events be slued around at right-angles to the rest, so that the time order of that area, as composed by its interior lines of strain and structure, runs parallel with a spatial order in its environment. It is conceivable, indeed, that a single whole human life should lie thwartwise of the manifold, with its belly plump in time, its birth at the east and its death in the west, and its conscious stream perhaps running along somebody's garden path."[9]

But Professor Williams does not really think that this 90° rotation is "conceivable." He has set it up as an Aunt Sally, as unlikely to occur as it is unlikely that "a human life be twisted not 90° but 180° from the normal temporal grain of the world," as in F. Scott Fitzgerald's story of Benjamin Button "who was born in the last stages of senility and got younger all his life till he died a dwindling embryo."[10]

It can thus be seen that Williams, in rejecting the concept of the passage of time, has at the same time unwittingly got rid of human intentionality along with the ideas of past, present, and future. In so doing, however, he imposes a serious restriction on the analysis of time,

[8] Donald Williams, (1951), "The Myth of Passage."
[9] Ibid.
[10] Ibid.

by limiting his conception of reality to the unidimensional temporal reality of the (3 + 1)-dimensional world of relativity. While this world may suffice for the work of the physicist and the natural sciences, it is entirely inappropriate for the human social world. It is a formulation which seriously impedes our understanding of human life by overly spatializing it. It illustrates the extent to which philosophy has been subordinated to the understanding of the natural sciences, and by and large omits the social sciences, or else tries to compress the social sciences into the physical conceptual world.

Moreover, once past, present, and future are recognized to be simultaneous regions in the field of human intention, there is no longer any need for the awkward and nonreferential concept of time's passage. All that is needed, in short, is to replace Professor Williams's "conscious stream along a garden path" by a different stream along a different path in another field—the unconscious stream of memory–perception–desire–anticipation along the path of coexisting past, present, and future in each person's mental field. It is this field which may now lead us to our two-dimensional temporal world.

Succession and Intention as Two Temporal Dimensions

Three main points have now been established which allow us to get through to our two-dimensional analysis of experienced time, which I am contrasting with both the single-dimensional viewpoint illustrated by Professor Williams and most physical scientists and the more-than-two-dimensional view illustrated by Heidegger (and by Sartre and others). The first point is that there is a well established philosophical precedent for the idea of the dimensionality of time to be a valid idea. The second point is that time *per se* is not directional.

The third point, which derives from the second and from the existence of the experience of succession, of earlier and later, is that neither time (that is self-evident) nor events are reversible in the world as known by human beings. Even if the universe were in a state of negative entropy and it changed to a state of positive entropy, it would not mean any such reversal as far as human knowledge if concerned. The statement of such a reversal is meaningless in the phenomenal world, whatever one's theory might be about the noumenal world. Earlier and later would still exist in human perception of succession, whatever new kinds of oddity might be occurring in the universe as it ran from states of maximum disorder to states of maximum order, or vice versa. What would not occur would be the reversal of world history so that what had already happened would begin to unhappen (Popper's fear of the elimination of the record of the occurrence of Hiroshima). Or even if such a wiping-out did occur, it would not be a true reversal of events or of time, since the "reversal" would have occurred later than the initial occurrence so far as any possible phenomenal human formulation is concerned, whatever the unknowable noumenal situation might be.

But we have also established that we do have a true directional sense when it comes to the past–present–future orientation of the field of unconscious mental activity, and of the preconscious awareness and conscious formulation and knowledge of that field of force. It lies in the directing of behavior in time in the true sense of temporal direction, namely, directing one's activity in the context of a directional organization of the mind toward a particular goal which exists only in the actual present as one part of that mental organization. I would point out yet again, however, that my analysis most definitely does not provide for an arrow of time, for a passage of time. We are dealing with the direction of human prediction and intention, the arrow of agency as Stuart Hampshire calls it,[11] expressed along the time axis of past, present, future. It is the felt arrow of our intended behavior directed toward achieving an aim and consequent satisfaction, and of our predictions about what will or may happen in the world.

There may be perceived, then, two distinct axes along which we organize our experience of time; the axis of succession, of the historical reconstruction of earlier and later, on the one hand, and the axis of intention, of simultaneous past–present–future, on the other.[12] These two experiences of time may be regarded as dimensions or coordinates of time, analogous to the three axes—the Cartesian coordinates—along which we organize our experience of space, and, like the Cartesian coordinates, they may be regarded conceptually as running at right angles to each other in sense. The axis of succession (see diagram 6.1) is the horizontal axis, running from earlier to later with temporal points tn, tn + 1, tn + 2, etc., at later and later points of succession. The axes of intention cut across the axis of succession, Mn, Pn, Gn; Mn + 1, Pn + 1, Gn + 1; Mn + 2, Pn + 2, Gn + 2, each of these axes representing the state of memory, perception, and goals of an individual at points n,tn + 1,tn + 2.

The units in which time is expressed are the same for both axes, just as the units of length are the same for all three spatial axes. These

[11] "[Doing things] 'with a view to' or 'in order to' are unavoidable idioms in giving the sense of motion of an action, the *arrow of agency* passing through the present and pointing forward in time." Stuart Hampshire, *op. cit.*, p. 73.

[12] Bergson expressed a like idea but left it undeveloped. He wrote: "Our perceptions actual and virtual, extend along two lines, the one horizontal, AB, which contains all simultaneous objects in space, the other vertical, CI, on which are ranged our successive recollections set out in time." H. Bergson, (1911), *Matter and Memory,* pp. 183–84.

Louis Mink also verged on this 2-dimensional analysis, as I described above.

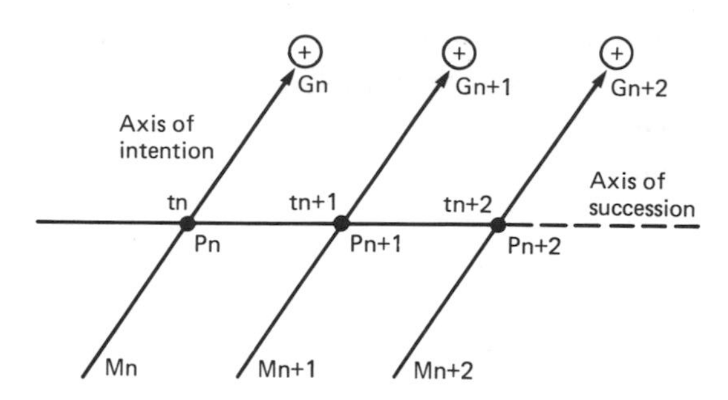

Diagram 6.1: Two dimensions of time.

units are the familiar calendar or clock-time units. They can be used both for statements about historical reconstruction and retrodiction, and for statements of predictions and intent.

We thus have two awarenesses of time in the individual—which cannot both be experienced consciously at the same time. When we are absorbed in our task our time sense runs from present toward future goal or desire, calling upon such memories of the past—experience and knowledge—as may help. We feel decidedly on the move into the future—our internal world future, not the physical time "future"—a future which we fully intend to create by actively controlling and transforming our perceived world.

While we are thus absorbed, however, our own activity is a process recordable on a clock as a succession without our necessarily being aware of it. For to be really absorbed in a task is "to forget all about time" (and indeed about space, for that matter). We may suddenly look up from the task, forget about it for the moment, and see what time it is (as a clock reading recorded along the axis of succession) and how much planned time we might have left, returning once more to the task with its future anticipation running at right angles to the axis of temporal succession.[13]

Kairos ***and*** **Chronos** ***as Dimensions***

This formulation of the directional and teleological sense of the temporal axis of intention, and nondirectional character of succession, can now be seen to be expressed in the Greek notions of *kairos* and *chronos* which were mentioned in Chapter I. We are also enabled to sharpen and clarify a useful distinction between the two—the one as the expression of intention and the other as the expression of succession.

[13] "The duration *wherein we see ourselves acting,* and in which it is useful that we should see ourselves, is a duration whose elements are dissociated and juxtaposed. The duration *wherein we act* is a duration wherein our states melt into each ther." Bergson, op. cit., pp. 243–44.

Let me recall that the idea of *chronos* is connected with the idea of succession—of a discontinuous succession of points measured by chronometer and recorded along a line, momentary and without purpose rather than momentous. By contrast, *kairos* was the time of the God of Opportunity, of personal action, of ends, time known by the content of intention and of prediction of what might arise.

The separation is strikingly similar to our two axes of time. The conceptions become effectively identical if, in place of the Greek conception of two different kinds of time, we substitute the conception of two dimensions of the one kind of time. In so doing I can formulate my central proposition in this way: the distinction between the time of the physicist and the time of the psychologist is not that between two different kinds of time; it is the distinction, on the one hand, between the one-dimensional time and the 4(3 + 1) dimensional world which suffices for the lifeless world of simple succession of physical things and events of the natural scientist; and, on the other hand, the two-dimensional time of succession plus intention and the 5(3 + 2) dimensional world which is necessary for the living world of wishes, of Desire in the Hegelian sense, of will and expectation, of the world of directional events of the life scientist and of predictive activity of the natural scientist.[14] We have not *chronos* as against *kairos*, but *chronos simplex* on the one hand and, on the other hand, *chronos plus kairos*, as right-angled coordinates of time.

The Representation of the Life-Space in Two Temporal Dimensions

I shall represent the moving life of the individual as a tube running along the axis of succession. We shall then be in a position to take slices across

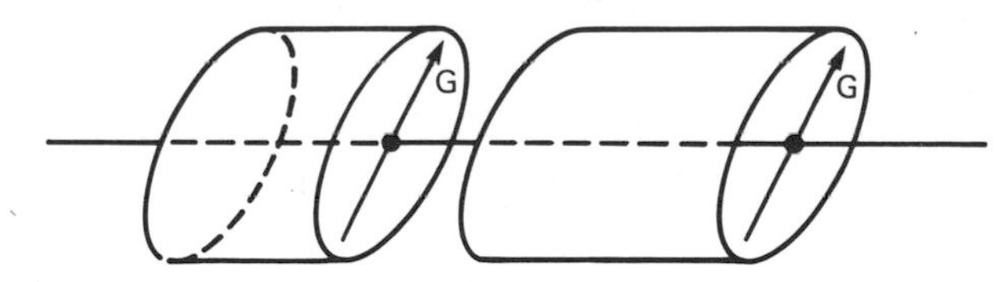

Diagram 6.2: Moving life-space of individual.

the tube, at right angles to the axis of succession, that is to say, along the axis of intention—and in so doing show the life-space of the individual at a series of points through time. In representing the life-space itself I

[14] Kelly, I think, is making this same point, but within the constraints of a single-dimension concept of time, when he writes: "There are some parts of the Universe which make a good deal of sense even when they are not viewed in the perspective of time. But there are other parts which make sense only when they are plotted along a time line. Life is one of the latter . . . . Life has to be seen in the perspective of time if it is to make any sense at all . . . . Life is characterized by its essential measurability in the dimension of time." This view is reinforced if time of intention is substituted for time in general. G. A. Kelly, (1963), *A Theory of Personality*, pp. 7 and 8.

am starting with the mode of construction developed by Kurt Lewin.[15] It will be necessary, however, to modify Lewin's form of representation for our present purpose, since in his diagrammatic forms he had difficulty in dealing with time and development. When he drew the individual's life-space in terms of structure of personality, time was excluded, as can be seen in the accompanying diagram 6.3.

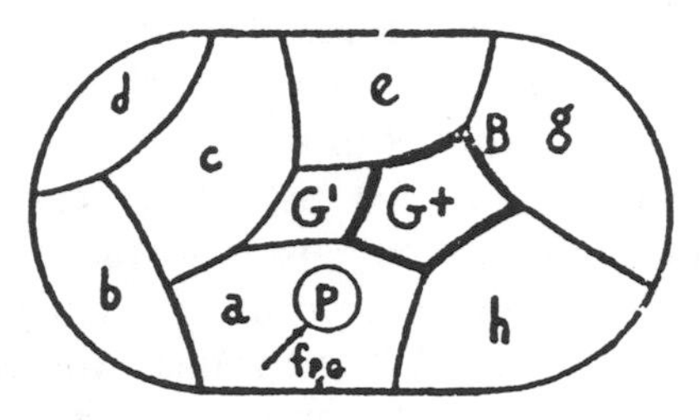

Field Representing the Conditions of Regression (According to Freud's Substitute Theory of Regression)
P = person; $G+$ = original goal; G' = substitute goal to which the subject regresses; B = obstacle between P and $G+$ (barrier); $a, b, c, \ldots$ regions of the life space; $f_{P,G}$ = force of the direction of the goal.

Diagram 6.3: Inner Life-Space
Source: Kurt Lewin, (1952), *Field Theory and Social Science,* p. 92

When, however, he introduced the element of time, as in his drawings of life-space showing locomotion through various regions, diagram 6.4, it is difficult to know whether this second diagram represents the

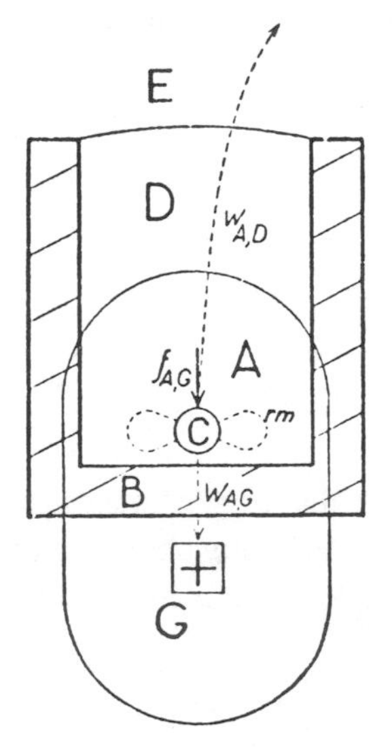

Diagram 6.4: Time represented as a locomotion: detour problem as seen by a child.
Source: Kurt Lewin, op. cit. p. 252.

[15] Kurt Lewin, (1935), *A Dynamic Theory of Personality.*

person's picture of a situation at a moment in time—in which case the time axis of succession is missing; or whether it represents a picture of an actual locomotion—in which case the time axis of succession is there but the time of intention is missing.

Very occasionally Lewin tried to represent both time of succession and time of intention, but in order to do so was forced to employ a rather awkward string of three balloons—past, present, and future—as in the accompanying diagram 6.5.

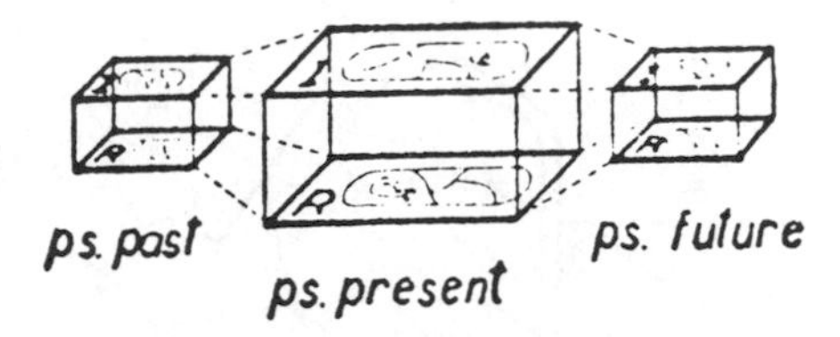

Diagram 6.5: Time represented as life-spaces in psychological past, present and future. Source: Kurt Lewin, op. cit., p. 246.

Reverting to our own diagram 6.2, it will be seen that an indefinite number of cross-sections can be taken at points along the axis of succession, representing the state of the individual at each temporal point. I shall show in a moment how this form of representation can be used to represent a total episode. But before doing this it will be necessary to systematize our representation of the main regions of the person in accord with our analysis of intention.

Diagram 6.6 shows one of the slices along the axis of intention, at a point t_1 on the axis of succession, in terms of a Lewinian life-space. We move from remembered past to perceived present to anticipated future across the slice from left to right; that is to say, all these time phases occur at the same temporal point of succession t_1, and together give the direction of intentionality. The slice is also divided into an upper and lower half, the upper half representing regions of the internal-world and the lower half the external-world regions. Finally, there is the inner region of active memory, perception, and anticipation, related to the particular goal-directed episode being considered. It is the focus, the foreground, the figure, within the background of less active or inactive memories, perceptions, and anticipations.

If we now focus upon the inner active region of the temporal space, we find six subdivisions or subregions. There is the subregion of immediate perception and desire: divided into the two further subregions of the inner psychological world of ideas and desires and of the outer world of things and events. Then there is the subregion of memories of events and experiences located in the psychological past: divided into the two further subregions of memories of events in the internal world, of thoughts, ideas, and conflicts, and of events perceived and remembered as occurring in the outside world. And finally there is the subregion of

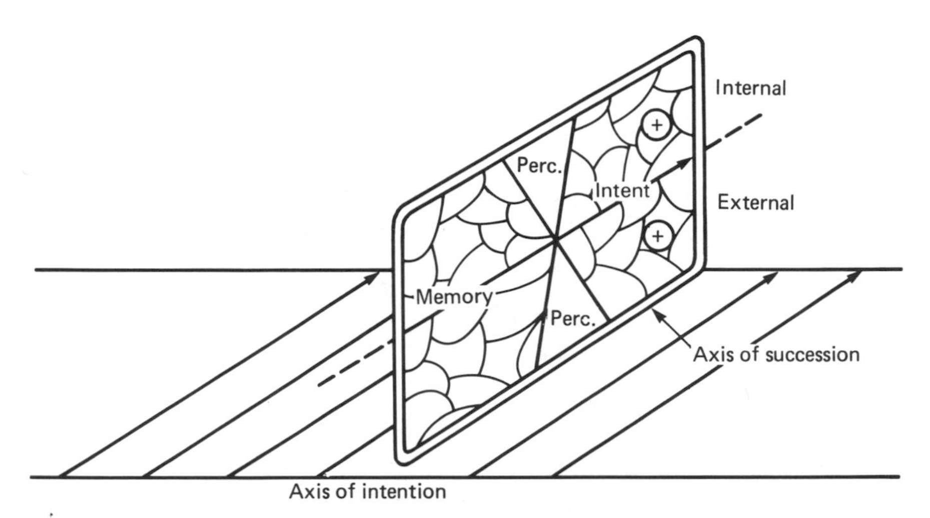

Diagram 6.6: Life-space represented as a section across the axis of succession.

anticipation of things to come, of things to be brought about by action stimulated and directed by intention: subdivided into the two further subregions of anticipation of inner psychological states and of circumstances to be brought about in the outside world.

The Teleology of Intention and Prediction

We may now note that it is the predictive and the purposive or goal-directed behavioral processes represented along the temporal axis of intention that have the quality of being directional. Here direction in connection with time takes on a precise meaning connected with a person willing himself to get to a goal in a certain time, directed toward that goal, pointing toward it, aiming at it, or his mental prediction that things are moving in a certain way. It is a *pointing toward something*. It is a going toward a *specified point*. It can be so only because of the existence of human prediction and intention which set that direction in terms of something that will happen or that is to be achieved.

It is a matter of some substance not to be deceived by the apparent directionality of the temporal axis of succession. Succession is in fact not directional, however much it may feel as though it is. It is neither symmetrical nor asymmetrical, reversible nor irreversible. All that it is possible to say about the temporal axis of succession is that retrospectively one thing can be seen to have succeeded another thing which recordings show to have occurred earlier—but not that one thing is moving toward another which does not yet exist. The temporal axis of succession is no more directional than the x, y, and z coordinates in geometry. It is a statement of a reconstruction of what has already succeeded what, and cannot be a statement of what *will* at some time in the future succeed what.

The prediction of a future event or the intention to bring about such a future event is a human mental construction; it can be represented only upon the axis of intention, and not upon the axis of succession. It is a probability statement at time ti_n of the likelihood of occurrence of an event at some time later than ti_n which would be similar in certain respects to the "future" event at ts_n it is an event whose only existence is in the mind of the predictor. Such a predicted "future" event is thus closely interwoven with the predictor's wishes and intentions. Indeed, it is to neutralize or rule out this influence of personal desires upon our predictions, this coexistence of prediction and intent, that experimental controls become necessary in scientific research.

Confusion arises because, in the absence of a two-dimensional analysis of time, the intentional goal time gets mixed up with the elapsed succession time. We thus say—on May 1st—that we are moving toward a particular event or place by May 3rd. May 3rd here is not time of succession—it cannot succeed May 1st because it has not yet happened. It exists as nothing more than an idea, a statement of intentional direction and of probability, in the present, on May 1st.

It is the confusion between the *two* temporal coordinates that leads to trouble. It is a confusion between succession and intention. A simple example is evolution. Evolutionary theory is a statement about succession. Its propositions have to do with the fact that particular species now in existence succeeded other species which are retrodicted to have existed and which in turn succeeded other species. From these propositions the relationships between existing species can also be posited.

But while biologists may try to use evolutionary theory to predict that certain species will emerge, such a mental construction does not mean some predetermined direction of development of species. Such predictions are directional only in the sense of their being an expression of the mental state of a biologist at a particular point in time: but they are nondirectional and nonteleological as far as the evolutionary processes themselves are concerned. These processes are matters of historical reconstruction represented along the axis of succession, and both nondirectional and nonteleological. It is not necessary to bring in God or any other "prime mover." Indeed, it was because Darwin was felt to have removed the hand of Godly intention from mankind that he was berated by some religious leaders, and not because of his linkage of man and ape.

From the point of view of our analysis, the concept of teleology, like the concept of direction, applies only to events which can be represented upon the axis of intention. That is to say, the behavior of living animal organisms[16] is teleological (goal-directed, purposeful, unconsciously directed), and directional (toward a particular goal), and the teleological and directional features (but not the succession features) are represented on the axis of intention. All physical processes (but not the scientist's mental predictions about those processes), by contrast, are nonteleological and nondirectional, and can be represented on the axis of succession alone.

A *succession* of events is not directional, it is not teleological. An *intention*, however, is both directional and teleological. It is the orientation of human intuition, the intentional process, which is directional, and not time *per se*.

[16] I am not including the tropism of plants.

PART FOUR

PURPOSEFUL ACTION AND TEMPORAL DOMAIN

CHAPTER SEVEN

The Goal-Directed Human Episode

Our exploration of the univocal concept of time and of the nature of the experience of time has led us to the realization that our way of organizing our world oscillates back and forth between two dimensions of time deployed at right angles to each other. There is the axis of succession threaded with an infinite number of discontinuous points along which can be mapped a recordable and dateable sequence of earlier and later events. Then there is the axis of intention abstractable as a single dateable slice cutting across the axis of succession, and on to which can be mapped the field of contemporaneous past, present, and future.

Each of the two dimensions of time is associated with a different cognitive modality. The axis of succession is experienced in terms of the atomistic modality, characteristic of conscious focused perception of things at a distance from other things simultaneously distributed in space as well as from other things distributed earlier or later in time. The axis of intention is experienced in terms of the less discriminated field modality, characteristic of preconscious awareness and unconscious sense, without focused things, without consciously known boundaries, but with a general sense of wholeness and intention and of unconscious contemporaneous field-of-force of past, present, and future directed toward a goal.

It is my expectation that this analysis of time as a univocal concept which can be treated as 1, 2, 3, 4 . . . n-dimensional, just as can space and the space–time plenum, may serve to release psychology and the social sciences from the invidious and inhibiting choice of being either "hard" sciences like physics or "soft" sciences more akin to the arts, literature, and history. It will allow them, and enable them, to become effective sciences in their own right and in their own way: with a quantitative rigor which includes objective equal–ratio–length scale measurement of properties of human entities, which gives no pride of place to physics and its like measurement of properties of material entities, and with a descriptive power which leaves the purposefulness and values of the meaningful human being and society intact.

This possibility of separating out a scientific method appropriate to human behavior from that appropriate to physics derives from the possibility of distinguishing the concept of a 4(3 + 1)-dimensional world from the concept of a 5(3 + 2)-dimensional world. The former concept has given a sufficient context or container for the analysis of physical events in space and time. The construction of the latter may give a sufficiently rich context to encompass and hold together the reality of purpose, goals, intentions, values, judgment, choice, and decision on the one hand, with the realities of place and time on the other hand, a conjunction which is essential if we are to be able successfully to characterize the fullness of living behavior in a scientific way.

We will examine in more detail the less familiar field modality, and its continuous unfolding in the form of coherent episodes of purposeful behavior.[1] These purposeful or goal-directed episodes have beginnings, middles, and endings. They are individual actions and social interactions replete with human desires and values, linked to the process of carrying out intentions to their point of completion. It will be suggested that it is by the abstraction of episodes by means of cuts along the temporal axis of succession that we constitute the basic "things"—the particulars—of the human sciences, just as the abstraction of physical objects by cuts out of spatial extension was the basic starting point for the natural sciences in ancient Greece.

In our analysis of the intentional episode it will also be observed how the unconscious mind oscillates between what may be called an orientating stance and an implementating stance—between planning, evaluation, and direction and goal-setting, on the one hand, and acting in the direction set and carrying out planned activity, on the other.

The continually shifting pattern of organization of mental activity along the temporal axis of intention is governed by the unique mental organization in each of us: our present perception of the present, pressed by our present organization of memories of the past, and pulled by the vacuum created by our present desires and by our present intentions and strivings to satisfy our desires whether in the immediate or in the longer-term future. The particular organization of memory, perception, desire, and intention in each person sets the limits of personal identity and of meaning and defines the individual self.

[1] I am indebted to discussions with Professor Daniel Miller for this emphasis upon the importance of episode as against static cross-section in the analysis of human activity. It is an orientation which occurs also in the writings of Roger Barker, Susanne Langer, and Stephen Pepper.

The Objective Study of Purposeful Behavior

General philosophy and the philosophy of science have managed to get by with a one-dimensional view of time—the temporal axis of succession—and a 4(3 + 1)-dimensional world, because they have been dominated by the problem of understanding the external physical world and natural sciences. With the exception of some few contemporary authors,[2] the requirements of the social sciences with respect to concepts of space–time—whether for the understanding of individual behavior, of social interaction, or of social institutions—have not by comparison figured very prominently.

Spatial concepts and the succession axis of time are, by and large, sufficient for the objectification of the external physical world—but even for this purpose they require an acceptance of the discomfort of oscillation between the two poles of the atomism-field duality. The degree of discomfort which can be stirred by this duality is reflected in the fact that Einstein himself spent the latter part of his life, when at Princeton, attempting to develop a unified theory based upon the idea of the inertial continuum, in order to get away not only from a simple atomistic concept of the world but also from the atomism-field theory dualism—"this disturbing dualism" as he called it in his autobiography.[3]

As we have shown, however, to limit ourselves to a one-dimensional view of time—the axis of succession—and to try to eliminate the atom-field dualism, will most certainly not do when it comes to considering events in the internal psychological world—including the world of unconscious processes—and the world of social systems and social interaction. We have suggested that the field-organized axis of intention is required as a second temporal dimension. But can it really be argued that this second axis is more than a play on words? Is it not, perhaps, an idea that is irrelevant because not much can be done with it? After all, the 4(3 + 1)-dimensional space–time coordinates are used for the practical task of locating spatial points in relation to one another in space and time. Are the two dimensions of time required for any such telling and essential purpose? The answer is yes. Their significance can be demonstrated in

[2] There are notable exceptions, of course, such as Popper, Northrop, Ryan, Kuhn, Braithwaite, who have included in their writings a concern with social science methodology as well as with the study of the material world. But nevertheless, even among such writers it might appear that the social sciences and psychology ride on the coat tails of the natural sciences, and not in their own firm right. The whole scene remains colored by the 4-D outlook toward the material world, and human meaning, purpose, and intent do not get much of a look in.

[3] He wrote: "Why not then *total* inertia? Then only field-energy would be left, and the particle would be merely an area of special density of field-energy. In that case one could hope to deduce the concept of the mass-point together with the equation of the motion of the particles from the field equations—the disturbing dualism would have been removed." *Albert Einstein, Philosopher Scientist,* p. 37.

the analysis of human reflection and social interaction. For human beings are purposeful and goal-directed, predictive, and probabilistic. They consider consequences in terms of if-then conditional outlooks ("if I do this, then I can probably make that happen"). And human behavior is regulated not only by continual analysis of feedback, but by the interplay between feedback information and intent. The understanding of all these features demands the second temporal dimension, the axis of intention.

It will be apparent that I am assuming a particular conception of human behavior. This conception is central to my argument that a two-dimensional analysis of time may give a more flexible and yielding foundation for the human outlook, and open up new areas to scientific thought and endeavor. I must make as clear as I can that conception of behavior.

In referring to the behavior of individuals and to the structure of social interaction, I am referring solely and exclusively to purposeful behavior, to behavior saturated with meaning and desire, impregnated with value and intent, directed toward goals, striving to achieve those goals, actively aimed toward intended outcomes. To understand behavior so defined—indeed even quite simply to observe it—requires not simply that we observe it in the limited sense of the physical locomotion of persons as physical or even as physiological entities, but that we observe it complete and in the round with the full assumptions we use and the interpretations we make about its meaning for those concerned and ourselves. It requires that we inform ourselves about the intentions of those persons, ask about their goals and desires, and about their assumptions about their interactions and relationships, listen to what they say, and try to grasp the meanings of their words and of their nonverbalized expressions and communications.[4] In short, from the point of view I am adopting, the statement that a person's arm moved is not in itself a description of behavior; it is a statement about the motion of an arm. For it to become a description of behavior would require a statement to the effect that "X tried to hit Y" or "X intended to shield the light from his eyes" or "X reached out his arm to pick up his glasses."

Of course, to define behavior in this way requires a sound definition of intention or purpose which enables it to be distinguished from the mere tropism of plants, or direct responsive activities of, say, unicellular organisms, or immediate reflex actions. I shall define intentional behavior

[4] This view is fundamental to psychoanalysis. It was developed most extensively in academic psychology by H.A. Murray at the Harvard Psychological Clinic, when he sought a comprehensive explanation of behavior in terms of the formulation of a system of individual Needs interacting with meaningful elements in the environment which he called Press. Murray also took time seriously in his emphasis upon development studies, an orientation which is now expressed in the Institute at Radcliffe College which carries his name, The Henry A. Murray Research Center. See, for example, his *Explorations in Personality*.

as behavior which shows evidence of deferment of gratification in the sense that there is a gap of however short duration between the experience of a lack or need and the action (behavior) to overcome it. Such gaps can be reported directly by human beings or inferred from the learning behavior of animals. They are evidence of the operation of organizing processes in the person.

That it is even necessary to take the trouble to define behavior in this way is a reflection of the fact that there exists an alternative definition, an alternative view, which obscures and holds at bay the full picture of behavior. That view has always exercised a powerful grip on human thought, and continues to do so. It is the mechanistic positivist view which appears in the sociology of Comte and its present-day derivatives, and in the so-called behavioristic theories in psychology. These views have one thing in common; they hold that in order to obtain a scientific understanding of human nature and of society, it is necessary to strip human behavior bare of all purpose and meaning. It is the purported vagueness and indefinable qualities of our assumptions about purposiveness, intention, meaning, which are held to be the culprits which obstruct truly scientific research in these fields, since they get away from the physical realities of physically definable behavior, and therefore—at least so it is held—get away from the hard reality and enter the realms of the artistic and the spiritual.

The current version of this positivist outlook is expressed in terms of the more mechanistic schools of cybernetics, information theory, decision analysis, and artificial intelligence studies.[5] These mechanistic views conceive of the human being and mental activity as a black box which connects an input stimulus on one side with a behavioral output on the other. Whether or not black boxes have intentions or purposes, whether they seek and take in information (intake rather than input) and create meanings, is left to one side. Life is seen as a more complex version of chess and other games, to be played with conscious logic, and with externally limited rules and a prescribed and given number of choices. The idea that it is the person who unconsciously creates and reformulates his goals, throwing up previously unknown and unforeseen

[5] I have already indicated in Chapter 5 that there are some cyberneticians and decision analysts who have demonstrated the shortcomings of this mechanistic approach. Pask, for example, argues that a central feature of real-life decision-making is the fact that individuals have to construct and modify their choices, the range of choice always being open-ended. Phillips, applying Ward Edwards's use of Bayesian statistics, has shown how it is possible to assist decision-makers by helping them to formulate and to become more aware of their own range of choices, and the probabilities they attach to them. Gordon Pask, (1969), "Strategy, Competence and Conversation as Determinants of Learning," pp. 250-67, in L.D. Phillips, (1973), *Bayesian Statistics for Social Scientists*.

possibilities and changing the rules as he goes along, is anathema, since it overthrows the apparent tidiness of the functioning of a human being as a machine (albeit a special kind of machine) and substitutes the apparent untidiness of behavior as it is in raw nature, seemingly inaccessible to systematic study and scientific knowledge.

I find it a gross abuse of the term "behavior" to apply it indiscriminately both to the operation of machines and to human activity as though there were no difference. Even in chess, human chess players—as compared with computers—can become bored, choose when or when not to play, overturn the chess board in a rage, change the rules and make up new games, and behave in an infinite number of ways which cannot be programmed into machines, because they include behaviors which had never even been thought of before the game; moreover—and above all—whatever they might choose to do, human beings can take responsibility for their actions. Nor can this feature of the open-ended plasticity of human life be duplicated by building random elements into our machines. For human thought and action is never random: it is always responsible, always determined—by the feelings of the moment, by current states of mind, by intentions, by desire and lust, by greed and gratitude, and even, for that matter, by mechanistic and positivist philosophies (as well as other philosophies, of course).

It might still seem, however, that behavioral episodes are less real, less tangible, than physical objects, because they cannot be observed all at once. They must be followed as they unfold, and then must be reconstructed retrospectively once they have reached their conclusion. But then so too is the observation of a physical object a process. No object can be observed all at once. The description of an object is also a retrospective reconstruction. The idea of an object that does not change, that remains fixed while we observe it, is a human fantasy, no doubt useful for a very limited range of studies in mechanics but a fantasy nevertheless. It is, moreover, a severely distorting fantasy, as Heisenberg has shown, even in the natural sciences, as soon as we get away from the very narrow confines of statics and mechanics.

I intend, therefore, to use the term "behavior" for human action replete with meaning, intention, purpose, responsibility, direction.[6] In order to reinforce this meaning, I shall sometimes use the phrase "pur-

[6] It might be objected that this particular usage in the meaning of behavior is unwarranted because it already has a well established meaning in behavioristic psychology and animal studies, and many behaviorists like, for reasons of their own, to believe that their work is freed from purpose. But, in fact, such studies posit purposefulness even in animals. Conditioned learning experiments, for example, require the psychologist to tune into the desires and dislikes of the subject—whether human or animal—and to establish a dialogue with or without the use of words in which the subject's purposes are not only assumed but understood.

poseful behavior,'' or ''meaningful behavior,'' or ''goal-directed behavior.'' But even where I use the term ''behavior'' on its own, I shall still mean purposive activity. I shall still be referring to action which cannot be known just by observing physical movements of people, but to action which includes both the interpretation of the meaning of what people can be seen to be doing and what people themselves convey as the meaning of what they are doing.[7]

It is this latter feature—interpretations and people's statements of the meaning of their own actions—which constitutes the great divide between the natural sciences on the one hand and psychology and the social sciences on the other. We can observe physical events—free-falling objects, for example—without having to worry about what the objects happen to feel either about falling or about gravitational attraction. The observation of two bodies falling in love, however, is a different matter, and so is the force of attraction between them. Because physical objects are intention-free, one temporal dimension will do for describing physical processes by dating their progress at successive times. Because human objects are intention-packed, two temporal dimensions are necessary—the one dimension to locate and describe the feeling, the other dimension to date the progress of the event.

It may seem, nevertheless, that physical objects and processes are more real than behavioral episodes precisely because physical objects do not fall in love, because they do not have intentions and desires which can only be inferred and not directly ''seen.'' Whereas an object can ''really'' be seen to fall, a falling in love might possibly be thought to be not more than a poetic metaphor. Such a view is tenable if one is ill at ease with human impulses; but if that is the case, then it is better to limit one's scientific curiosity to electrical impulses rather than to study human existence.

Moreover, balancing the fact that the physicist does not have to bother about the utterances of his material is the fact that the social scientist can elicit the cooperation of his subject, can ask him what is going on, can listen to his talk—and the replies and comments constitute objective facts. Of course it can then be argued that all such statements are difficult to assess. They may be rationalizations, or distortions, or evasions, or downright lies. And even when a person is not consciously trying to be secretive, there will always be great discrepancies between his conscious knowledge of his motives and intentions and the uncon-

[7] This view coincides with one of the modern trends in philosophy which is concerned with the person-in-action. MacMurray is an outstanding exponent of this approach in his *The Self as Agent*, (1957) and *Persons in Relation* (1959). So also are Stephen Pepper, (1970), *The Sources of Value;* and Susanne K. Langer, (1967), *Mind: An Essay on Human Feeling*.

scious determination and direction of his behavior (discrepancies, indeed, which I used to support my whole argument in Chapter 5).

But—and here is the core of the issue—the solution to the problem lies not in dehumanizing behavior by banning meaning, intentions, values, responsibilities, and purposes from scientific study, by deeming them unworthy of the scientist's efforts, by outlawing them to the artistic hinterlands. It lies in bringing them into the center of the arena, focusing upon them, putting the spotlight upon them, making them an integral part of the study. In adopting this viewpoint, I would nevertheless agree with possible critics who would say that to establish the nature of people's intentions may often be a very difficult and uncertain business. Of course it is, even for the people themselves. It must always be a matter of successive approximation and interpretation. But so long as we are not afraid to make interpretations and to proceed by successive approximation, and are intent on testing and retesting our conclusions as we go in order to ensure that we can eliminate errors, then there is no need to eschew such procedures.[8] Indeed, far from being a cause of disquiet, it is precisely this uncertainty about human motivation, intentions, values, meanings, relationships, comings together, social structurings which makes them of interest to study; it is the spice of the study, the intriguing part, the human part, the living part, the social and the political and the economic part, the part about why people are like they are and do what they do.

I shall elaborate to some extent upon this very substantial problem of determining purposes, values, intentions, desires, expectations, goals. But my main point will be that, in order to tackle the problem at all, it is necessary in the first place that these concepts be given room to breathe in an adequate temporal framework (one which includes an axis of intention) and freed from imprisonment in a 4-dimensional physical cell with room only for an atomic immobile single time axis of succession.

In the natural sciences it is quite possible to study many aspects of the physical world in terms of a static 3-dimensional spatial abstraction as though the world were standing still, and to deal with dynamics, with process, by the addition of the fourth dimension of time. In the human

[8] Gilbert Ryle has put the matter very succinctly, in terms which echo the views of Karl Popper, in a note on epistemology: "The experience which is omitted from the theories of the empiricists is the experience which is omitted from the theories of the rationalists. Craving for something to avert the possibility of mistakes, the one finds its haven of safety in uncorrupted sense-impressions, the other in uncorrupted ratiocination. But the successful investigator is he who has made sure, not he who has remained in safety. Where mistakes are possible, the avoidance, detection and correction of them is possible. Knowledge comes not by some immunization against the chance of error, but by precautions against possible errors—and we learn what precautions to take by experience i.e. training and practice. It is the expert, not the innocent, who knows." "Epistemology" (1966).

sciences there are no statics. Nor are there any intention-free 4-dimensional dynamics. Living phenomena are moving by their very nature, and moving with directional intent. If they are not purposefully moving they are dead. In order to study them, therefore, our abstractions must be in terms of events defined by purpose, of episodes with a beginning, a middle, and an end which have meaning for the persons involved. The idea of purposeful event is implicit in the idea of behavior, of action; and the idea of episode is implicit in the idea of intentionality and goal-directedness of that behavior and action.

The Goal-Directed or Purposive Episode

Behavior comes in temporal episodes. Episodes are abstractions from the space–time continuum, in the same way that physical things are abstractions. They are the fundamental things which we can study and observe in the human sciences. They are the basic building blocks for those sciences. They are as real as the material objects, substances, and atomic and subatomic particles of the natural sciences. A behavioral episode is tangible, encompassable, definable, observable, describable, just as is any event or process in physics or chemistry.[9] Any completed episode is bounded by two recordable points on the temporal axis of succession. And during the episode there occurs a continuous changing of the structure and content—the patterning–of the unconscious field of memory–perception–intention of the participants as mapped along the temporal axis of intention.

The three terms—goal-directed, problem-solving, and purposive—are used extensively in psychology to refer to the type of episode we are considering, but usually in connection with the idea of goal-directed, problem-solving, or purposive behavior in general. I wish to use episode rather than behavior in general, because I wish to emphasize a variety of aspects of the episode—how it begins, how it runs, and how it finishes, the amount of time committed or allowed for reaching the goal, the organization of activities to bring the end into being, the sense of urgency if the end seems to be elusive, the amount of time actually taken, and so on.[10] An episode has an identifiable and analyzable structure as compared with purposive behavior in general.

[9] Kelly has put this point well: "The subject of psychology is assumed at the outset to be a process . . . . the person is not an object which is temporarily in a moving state but is himself a form of motion." G.A. Kelly, (1963), *A Theory of Personality*, pp. 47 and 48.

[10] As Kelly puts the matter: "Only when man attunes his ear to recurrent themes in the monotonous flow [of undifferentiated process] does his universe begin to make sense to him. Like a musician, he must phrase his experience in order to make sense of it. The phrases are distinguished events. The separation of events is what man produces for himself when he decides to chop time up to manageable lengths." Ibid., p. 52.

In outline, a goal-directed, problem-solving, or purposive episode has the following characteristics. Here they are summarized, and then detailed in turn. Out of the more detailed analysis some of the outstanding features of the structure of experienced time will emerge.

A goal-directed episode begins with a person with a sense of something particular which must be done. It may be stimulated from outside by some thing or occurrence which attracts our attention and which we then choose to pay attention to—by a chance encounter which must be followed through or by an invitation to take part in some event, or by an instruction to carry out an assigned task. Or it may be self-initiated, beginning with a *lack,* a felt need, a gap, a sense of something missing, an absence, and a wish that this gap or lack or absence did not exist. It is a feeling of discomfort, whether stirred by a need or by a sense of obligation or duty, and may range from a vaguely felt sense of lacking something to the most powerful feelings of grief over the loss of someone or something which, however much desired, can never be replaced.

However the episode begins, whether from some intake of an external stimulus or from an internal arousal, the individual experiences a need, or lack, in the sense of an episode to be traversed, a goal to be reached, an event to be completed. It might even be an unpleasant event, perhaps a chance encounter which he wants to complete as quickly as possible. But pleasant or unpleasant, the individual is in an unsettled situation—there is an immediate sense of lack, of unfinished business, and there is a goal to be achieved to fill that lack.

This sense of lack next begins to take shape as a *desire*—the most powerful of human feelings, the prime mover of men and women, the moving spirit of the psychological and social world, the source of all human activity, of creativity, of conflict, of cooperation—the Desire emphasized and written with a capital D by Hegel.[11] Desire takes the form of wishing or willing the existence or the completion of *something* which can replace the lack or somehow complete or close the episode—the world of will and idea of Schopenhauer.[12] It is something to be possessed for oneself or done for someone else. It might exist and is somehow to be obtained, or it might have to be found, or even constructed or created, or simply experienced, seen, listened to, or gone through.

This something at the initial stage of desiring is a *goal image;* it is an idea of what might satisfy the desire, fill the gap, turn a lack into a feeling of satiety, complete a chosen or required or imposed activity. It is not as yet the real thing, but rather an idea of what the real thing is

[11] Hegel, (1807), *The Phenomenology of Mind.*

[12] Arthur Schopenhauer, (1966), *The World as Will and Representation.*

(if it is known to exist) or might be (if it does not yet exist or at least is not known to exist).

Out of this goal image, *orientational* and *exploratory* behavior begins, which might range over simple quiescent thought and reflection, physical looking about, research and study, or any of a series of other activities which can help to form a *plan* as to how a satisfactory *goal object* can be obtained which will be true in the sense of corresponding sufficiently to the goal image to fill the gap.

This plan must have as one of its components, if it is to be a realistic plan, the availability of *resource objects;* that is to say, resources in the form of things, people, ideas, which exist objectively and which could be used as satisfactory goal objects either as they are or as they might be if suitably transformed.

The remaining problem is to traverse the *path* which has been planned toward the attainment of the goal object, overcoming such *obstacles*—expected or unexpected—as may appear on the way, until the goal object is created or otherwise obtained, the lack fulfilled, and the end state of quiescence achieved; or else the goal object is modified—or else it is abandoned and failure is experienced.

This description of the general features of a behavioral episode may serve to highlight the definition of the active present, as against the immediate present and the existential present which I have described in Chapter 6. The active present is composed of all the behavioral episodes in which someone is engaged—all the intention-filled trajectories from their beginning to their final end state if achieved, or to a sense of failure if not achieved. It is how real life is experienced, and in particular organized. We arrange our activities, our plans, our hopes and expectations, our choices and desires around episodes; we bring various regions of our past experience into active play in relation to the current episodes in which we are involved. We set our priorities, our urgencies, our first things first, in terms of episodes. In short, we organize and actually live our lives in purposive episodes, in intentional trajectories, which are present, in existence, until they have reached their mark or have been dropped or transformed. That is the fullness of our active present: it is our sense of "present-ness" extended to its widest active limit.

The concepts of the purposive episode and the active present will thus become the central concepts around which my argument will revolve. It will be useful to note at this point, as an indicator of our path, that each person lives with more than one episode at a time (except perhaps the mentally handicapped, who are identifiable as mentally handicapped because of their inability to do so) and that one particular episode is dominant in the sense of being in the foreground at any given moment, but there are many other episodes in the background, all being actively pursued but perhaps momentarily in suspense, or idling, or awaiting their chance to take over the forefront.

Temporal Location of a Goal

The 2-dimensional temporal analysis is necessary if we are to locate human goals or intentions, and to follow their vicissitudes through a temporal succession. It is necessary to be able to specify the date on which a particular goal or prediction existed (the axis of succession), and the time by which it was intended or predicted that the outcome would result (axis of intention). We can then observe the progress of the carrying out of the intention or the actual unfolding of the predicted events, through to success or failure in achieving the goal. It must be noted—and this point is extremely important—that the goal itself is not a static entity, is itself a process undergoing continual modification during the episode as a result of the encountering of obstacles or changes in desire.

Thus, for example, one might say that at 12 noon on May 16, 1974, a person decided that he would trade in his car for a new one, and that he would attempt to achieve the change by June 30 in order to have the new car before his summer vacation which begins at the end of July. Such a goal locates temporally on the axis of succession at May 16, and on the axis of intention at 45 days planned for completion. He sets to work in intermittent bursts of activity punctuated by periods of reflection as he engages in other activities, and collects information on cars, and by the end of three weeks has test-driven four models. This progression is shown by the curve from point *a* (May 16—45 days) to point *b* (June 6—24 days).

He then discovers that there may be some difficulties because of extended delivery dates (the expression, note, of someone else's intention,

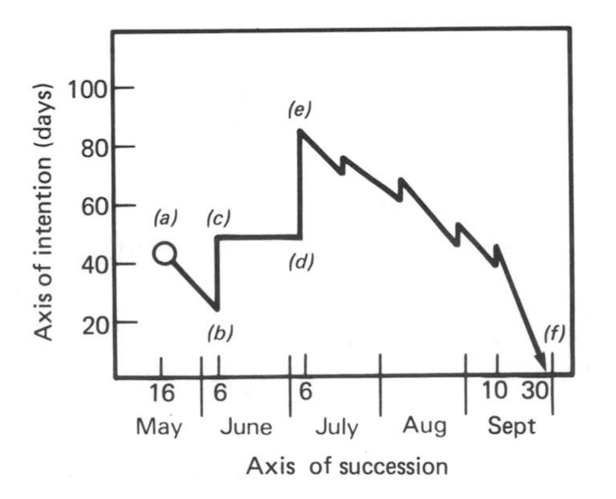

Diagram 7.1

and certainly not a predestined fact), so he decides to give himself an additional four weeks to find what he wants, as shown at point *c* (June 6—50 days). Another four weeks go by and, despite diligent searching, he finds that he will be unable to get delivery of a satisfactory car before the end of July, a situation plotted at point *d* (July 6—14 days).

In view of these difficulties he decides to give himself much more time, until after he gets back from his vacation. Meanwhile he will continue to think about the matter during his vacation, keep his eye open for new cars, and be in a position to choose and get delivery on a new car by the end of September, a revision of plan which is plotted at point *e* (July 6—85 days). And let us suppose he succeeds in ordering a car by early September, and by judiciously pressing the dealer gets delivery at the end of the month. That final outcome is plotted at point *f* (September 30—0 days) which shows the goal achieved.

The diagram illustrates the shifts in intention, with changes in the target completion time during the activity, on specifiable dates. The process could not be followed in terms solely of temporal succession or solely of temporal intent. Both axes are required to locate the goal at each stage and to describe each episode. It is to be noted also that as long as the person publicly states and shares the modifications to his intention, then the process is objective and recordable. It is an objectified internal event or process, as against the objectified external events of the natural sciences.

In effect, the two-dimensional time chart gives the history or development of a goal through time. It illustrates the important fact that a goal is not a static thing; it is a living changing entity, as are all social and psychological things (as we shall demonstrate). As the individual, having established a goal, progresses toward it, he will modify that goal in terms either of its content, or of its targeted time of completion, or both. The goal will terminate when the person either achieves it or else decides to abandon it.

Thus in diagram 7.2 are illustrated various goal sequences:

(a) the intention to have lunch in an hour's time;
(b) an episode whose target completion time was steadily increased and the goal eventually abandoned;
(c) an episode carried out in a series of regular moves and the goal achieved on target;
(d) a goal achieved more quickly than initially targeted.

The need for a two-dimensional coordinate system becomes apparent if the attempt is made to illustrate these sequences by mapping them solely along the commonly used dimension of succession—the time or-

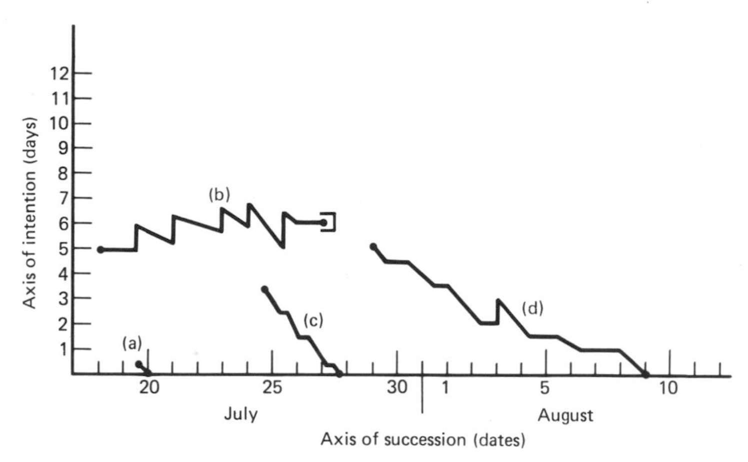

Diagram 7.2

dinarily designated by the symbol "t": it is impossible under this condition to show the interaction of intention and succession.

Action as Orientation and Implementation

I have so far concentrated on the person in action proceeding toward a goal. But activity of this kind is preceded by an orientation phase, and punctuated by reorientation phases in which the person holds fire, takes stock, decides what to do, and then starts up again. It is in these orientation phases that the organism consciously sets course, sets direction, points itself in that direction, and then releases its behavioral brakes and moves on.[13]

This process of intermittent orientation and reorientation is represented on the accompanying diagram (7.3). Each of the round dots on the goal graphs indicates a point at which the person in effect stops, considers or reconsiders the position, and leaves intact the goal, or changes it—either by shortening or lengthening the target time or by changing the content of the goal in some way. In order to achieve such a confirmation or modification of goal, the individual must take stock; there is a suspension of action and progress, a marking time, while memories of the past—both immediate and longer-term—are used to inform the present in relation to the desired goal in the future. Another way of putting it is that the individual is continually making feedback evaluations in the context of his past experience and knowledge and in relation to his intentions.

[13] This process has been expressed by Tolman as, "To be conscious is to hold up and delay, in order to enhance, to limn in some area or aspect of a position-field." E.C. Tolman, (1967), *Purposive Behaviour in Animals and Men,* p. 208.

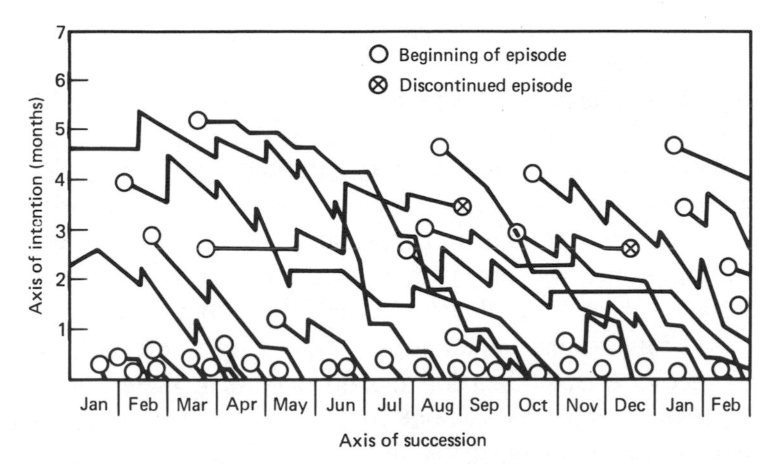

Diagram 7.3

Orientation goes on continuously as a person acts; it provides a continuously adjusting directional frame. But this continuous adjustment process can be distinguished from the point where doing is itself explicitly discontinued for a time, and replaced by reflection: as when a motorist stops to look at his road map and reconsider his route; or a writer reconsiders the plan of his story; or a machine operator revises his method of operation because of unexpected difficulties with the quality of the particular material he is machining; or a salesman pauses before entering to see a client for a last-minute thinking-over of the way he has decided to approach the client.

In the same way implementing action never ceases completely. Even during an explicitly organized orientation phase, there will be tiny motor movements, symbolic, restrained, practice movements, as-if movements. Such movements are an essential part of orientational mental activity. The organism functions as a whole, as a sensory-motor, thought, feeling, acting unity. Lop off any one activity completely and the rest also will die.

Because of the fact that during an orientation phase there is an explicit withdrawal from implementing action, and a retreat toward thought and explicit verbal formulation of plans on how to implement, it is possible for individuals to combine to make orientational decisions. They can get together to talk together and to arrive at decisions together. This kind of orientational activity is the kind for which we set up committees, councils, boards, parliaments—bodies whose functions are exclusively concerned with establishing the goals and the context for the actions of those represented by such bodies. It is this orientation and direction which settle for the group concerned the social meaning of its living and working together.

Orientation phases alternate with implementation phases. They may range from the briefest and most fleeting looking up from a task and taking stock—as does a concert pianist during a performance—to the establishment of more or less full-time institutions to carry out orientation functions—as does a parliament. People do not become inactive as against active at this stage. The organism is always active. But during an orientation phase there is an almost total inhibition of the motoric, or, as in the case of the pianist, motor activity is sent flying off on its own for the split-second orientational looking up (anything longer than a split second and the performance would disintegrate). The activity of the organism is almost wholly mental activity with minimal involvement of the voluntary musculature. Then when direction becomes clear again, there is a release of motor inhibition and the organism begins to move purposefully along the lines set by its own orientational activity.

This formulation of orientational and implementary action is solely in terms of meaningful and purposeful behavior and mental activity of the active organism. It is the total active, seeking, exploring organism which is involved in the orientational activity with its motoric inhibition. It is equally the total organism which carries through the meaningful actions directed by the orientational plan, employing an exceedingly subtle mechanism of highly selective inhibition and release. There is no need to suppose a mind causing a body to move. Nor is there any need to talk about a body containing a mind. We require only a living organism, continuously active from birth, exploring its environment, with selective inhibition of the motoric to achieve movement in line with intent toward a goal. In Lewin's term, B-fPE—behavior is a function of the person and the environment.[14]

It is only if we abstract a mindless body, alive or dead, and a detached bodyless insubstantial mind that we then find ourselves saddled wih the body–mind problem, and with the problem of how to link body and mind via dualism, or interactionism, or psycho-physical parallelism, or some other mechanism. But why anyone should wish to abstract a mindless body and somehow get it moving by a bodyless mind is difficult to understand. Some philosophical problems really are derived from extraordinary fantasies. Mindless bodies are dead bodies, and will be moved not by a mind but ordinarily by an undertaker or an anatomist. And as for a bodyless mind which is somehow located in a body but not an

[14] K. Lewin (1935), *A Dynamic Theory of Personality.*

inseparable quality of a living organism, that idea takes us directly into poltergeist theory.

Orientation, intention, unconscious mental activity, motor behavior are all constant features of all living organisms and are what we mean by a living organism. The mental activity is not located in a brain, and the behavior in efferent nerves and muscles. They are functions of the total living organism, not parts of that organism. I shall discuss this point further in Chapter 12 to show that all human phenomena require to be defined as intentional processes and not as atomistic spatialized things or processes—behaving persons engaged in intentional events, and not behavioristic fictions called bodies (other than dead ones) and minds.

In short, behavior is directed by the active orientation of the person and given meaning by that orientational context. A part of this orientation is the person's determination of the time by which a goal is to be achieved. It is his decisions about time which give him his sense of urgency and of priority for whatever he is doing, and which set the frame within which episodes are organized and come into existence.

Analysis of a Behavioral Episode

This constant changing and development of memories, perceptions, desires, intentions, and expectations, and the oscillations between orientation and implementation can be demonstrated in any episode in which we work toward a goal. I shall illustrate the process with an example—that of a person sailing a small boat to a particular port across a wide bay.

He goes into an orientation phase in order to plan his course across the bay. He sizes up the external situation by direct observation, charts, instruments, and other information—wind direction and strength, time and tide, local currents, position of rocks and wrecks, lighthouses, likely steamer traffic, weather forecast, and other relevant factors. At the same time there gnaw away at the back of his mind a number of memories—of having made the crossing twice before and of running into trouble because of tricky crosswinds and a strong ebb tide, but also of how pleasant it was finally to make port and to get back to the comfort of the attractive hotel where he was staying; and these memories are suffused by deeper-lying memories of being confronted by other threatening situations which pour through into the present situation and provoke vague feelings of superfluous anxiety. There would of course be an infinity of additional percepts, memories, and anticipations—from the arousal of infant memories of frustration and gratification to concerns about other current circumstances and plans—but these few data will suffice for our illustration.

There are an open-ended number of choices which he might construct by imposing his own knowledge, awareness, and hopes, expectations and desires. There is an open-ended quantity of information available to his conscious focus, to his peripheral awareness, and to his unconscious sense of the situation. He might stay where he is; or cross the bay; or land somewhere else. And having decided to sail, there are innumerable routes and tactics by which he might do so; whatever he decides, he could consciously explain. But his explanation would fall far short of why he intended what he did—far short of giving any picture whatever of the richness of the unconscious mental field of his memories and experiences, his organization of the way he chose to perceive his situation, and the particular balance and intensity of the complex of desires, hopes, stimulation by danger, symbolic meanings of the sea and sailing, sense of challenge, self-realization, defense mechanisms, public and private identities, internal objects, which make him the particular person he is, and which ensure that whatever he does will be different from whatever someone else would have done, or indeed, what he himself would have done at an earlier or later date.

It is out of such considerations that the sailor formulates the course he proposes to follow, and he sets off. It is 1300 hours, and he plans to make the crossing in about four to five hours.

After about two hours' sailing, the wind has strengthened unexpectedly and the tide has begun to ebb strongly so that the original plan becomes untenable. It would be impossible to reach port at the targeted time. The sailor revises his plan, senses his overall situation in his own personal way, and decides on a more circuitous route. He calculates that he can reach his intended destination an hour later than he anticipated—by 1830 hours instead of 1730 hours. He does not want to be later than that because by then it will be getting too dark for safe navigation.

Things go from bad to worse, however. The wind does not let up but veers to an offshore direction, making it exceedingly unlikely that he can reach port before dark. By 1600 hours the sailor therefore changes his plan completely, and decides to head for a more accessible shelter—a small fishing village below the headland—which he reaches by 1800 hours and where he moors his boat for the night, with the intention of sailing to his home port the next day.

To represent these changes in intention—changes in both goals and planned routes to the goals—requires a series of temporal space diagrams. They cannot be represented on one life-space representation of a locomotion as Lewin tried to do, because the changes in the goal itself cannot be represented. There is required an analysis of the situation at a series of points on the temporal axis of succession, the analyses themselves being carried out at each point in relation to the axis of intention.

The episode itself has occurred in time of succession, readings for which can be taken at various stages of the process. But the person's experience of the process—if, as we may assume, he has been assiduously preoccupied with overcoming his navigational problems—is of a time structure in which memory (past), perception (present), and desire (future) are actively integrated at every moment of the process, but conceptually cutting at right angles across the continuum of time of succession. As the process flows, the relationship between the present past, the present present, and the present future is constantly changing; the resource obiects become modified; the goal image is changed in adapting to the changing circumstances; experience of sailing is itself increased as the process goes on, so that the sailor's memory field is gradually enriched.

Thus, the present is a flowing amalgam of the person's present experience of where he is; of his present experience of the future as where he desires to be getting; and his present experience of the past, the experience he is using, including what he has already done—all of which exists as a unified complex field on an axis at right angles to the time coordinate of succession with its clock which ticks away without the person's necessarily being conscious of it.

As far as the memory of past experience is concerned, not the whole is called forth—only that part of the memory system relevant to sailing is brought into focus; but the whole of the memory of past experience is potentially there, acting both as context and as additional store of information should unexpected obstacles arise. Similarly, not the whole of a person's aspirations for the future are there in focus. Only those matters concerned with his aspiration to sail to his destination—and perhaps other more complex features of his hopes and desires such as substitute gratifications, compulsions, sublimations, and other unconscious meanings of sailing. But most of his ambitions will be held in suspense or may be actually dormant for the moment, some to be resuscitated when he has reached his destination and he moves on to the next order of business in his life.

This example may illustrate the two-dimensionality of our perceptual organization of time. Time of succession is a series of time readings—it does not have direction in the space–time manifold. It has no past or future. It is a series of static abstractions of points in time. In time of succession there is no past, and no future, and no significant meaning to the present. It can be socialized, in the sense of being expressed in symbolic form of clock time, and the clock readings shared with others.

Past, present, and future by contrast are conditions of mind of P as he pursues his aspirations (future), using his memories of experience (past), and his perceptions of both inner and outer world (present). Past, present, and future are all simultaneously and continuously together as one integrated field in his mind as he orients and implements in the course of this sailing episode. They constitute the changing content of the person's outlook mapped onto the temporal axis of intention.

The episode started with the sailor's not being where he desired to be, which led to the intention to get to port, and ended with consummation. The *form of time* lies in the fullness of the two-dimensional temporal feature of the sailing episode—the changing form of the field of the life-space of the individual throughout the episode as described along the temporal axis of intention, and the dating of these changes and the progress of the episode along the time axis of succession.

Zuckerkandl and the Musical Time Concept

This notion of the purposive episode as one whole reorganizing thing held in suspension in the present with the beginning, its unfolding, and its approaching termination, all there together in dynamic connection, can be seen in pure culture by listening to music. It will be worthwhile to consider for a moment the work of the musicologist Victor Zuckerkandl.[15] He wrote with great sensitivity on both the sense of time and the sense of space as a dynamic field of force rather than as a static extension of points at a distance—in connection with our experience of melody, rhythm, and meter in music, and our experience of pattern, tension, and harmony in a painting.

Zuckerkandl distinguishes between what he calls the physical time concept and the musical time concept, setting out the following schematic comparison.[16]

Physical Time Concept	**Musical Time Concept**
Time is *order, form* of experience	Time is *content* of experience
Time *measures* events	Time *produces* events
Time is divisible into equal parts	Time knows no equality of parts
Time is perpetual transience	Time knows nothing of transience

This distinction between physical time and musical time is strikingly close to our distinction between the temporal axis of succession and the temporal axis of intention. Zuckerkandl, however, does not keep both

[15] See Victor Zuckerkandl, (1956), *Sound and Symbol: Music and the External World.*
[16] Ibid., p. 202.

patterns of time as two separate coordinates, as I would wish to do. Rather he follows Heidegger into a rejection of physical time as having anything at all to do with time. He attempts to absorb all time, using his musical time as the model, into Bergsonian durée. Nevertheless, I have found in his analysis of music in terms of field theory—that is to say, of fields of force extended in time—an endlessly fascinating source of confirmation of the importance of the episode abstracted from the space–time manifold, and discontinuous with it, but having the property of continuousness, of field, within itself.

"Hearing music," Zuckerkandl writes,[17] "we oscillate with its metric wave. Each tone falls on a particular phase of this wave; each phase of this wave imparts to the tone that falls on it—and through the tone, to the auditor—its particular *directional* [my italics] impulse . . . . I know that I feel that, with this tone, I have reached the wave crest and at the same time have been carried beyond it, into a new wave cycle . . .. . I am able to hear directly from the tone—and from the rest—in what part of the entire measure I am at the moment."

This experience of wholeness is thus related to the concept of fields of force. "The remarkable fact that . . . the whole is in some manner present in the part—to this fact our thinking seeks to do justice by the *field concept*." The field is illustrated by sentences in language, in which the meanings of the word parts are given by the direction of the whole, and the meaning of the whole given by the movement and duration of the parts.

The whole field in hearing music, Zuckerkandl notes, is "given by tones of definite durations and a listener." Then, in order to explain how this can be, he argues: "The answer to the question 'of *what* are metre and rhythm the effects?' can, then, only be that they are the effects of the mere passing of time in the tones, of their temporality. Because tones have duration, because time elapses in them, and for no other reason, we have the rhythm of our music. Only process through time can be the agent and source of the forces active in metre and rhythm."

In his analysis, Zuckerkandl tends to reify time, "mere time," as somehow something, albeit a force, which makes for sound and music. That he does not need to do so, however, can be shown simply by not forgetting, as he too quickly does, that he has "a listener" as part of his field. The moment we remember to include the listener we add to the total field the sole feature which is capable of giving the "directional impulse" to the sequence of tones and rests. And for this directionality to occur, there must be at least a *minimal* intention on the part of the

[17] Ibid., pp. 204 to 208.

listener to listen. Listening to music is an active process of attention—of excluding the rest of the world of sound.

In listening to music there is in fact a complex human intentionality and directional impulse of an interesting kind. It takes us into some of the basic features of interpersonal relationships. There is in the first place the intentionality and directional impulse of the composer who creates and records a repeatable human episode. There is then the intentionality of the performer (or performers) who imparts his interpretive directional impulse to the composition. And only then do we have the intentionality and directional impulse of the listener tuning in, to greater or lesser extent, more or less successfully, more or less comprehendingly, more or less rapt and enraptured, more or less attentively, to the tones and rests offered in the auditory surround of his behavioral field.

Listening to a musical performance, then, is an example of a human episode—with a beginning, a middle, and an end. For someone with the intention to listen, with the interest, the competence, and the desire to do so, with sufficient experience to call upon—or to press upon the present—hearing a piece of music is *one* continuous event; it is present as one coherent actively present episode, incomplete until the final climax resolves and redistributes the whole and completes each and every part.

We can thus translate Zuckerkandl's concept of time as force, into the idea of human intention as temporally expressed force acting in a direction set by goals and intentions. That such a translation is not far removed from his own thinking is shown in his interesting concept of "auditory space." By auditory space he means the sense we have of a spatial wholeness that accompanies our experience of a musical episode. "With the sounding of the tones, a different order in the 'whence of encounter and the where of relation' opens, and now does for tones and tonal motions what the order of places does for bodies in visual space: it keeps separate things apart and relates them to one another, and thus makes it possible for ordered creations of a higher kind to originate." [18]

The source of this experience of auditory space Zuckerkandl connects with the biologist Von Uexküll's concept of "plan" as an essential notion in understanding living biological events in time as with auditory events, rather than in space as with physical things distributed at a distance. Von Uexküll's concept of plan expresses the biological foundation of human intentionality.[19]

Not content with music, Zuckerkandl takes his powerful grasp of dynamic fields of force into apparently physical space itself: "space as

[18] Ibid., pp. 332–33.

[19] This connection is tied in with the fact that biology has always been exercised by notions of vitalism and teleology. Jacob von Uexküll, (1962), *Biology*.

force,'' as he dramatically calls it, as compared with the ''space as place'' of the ordinary everyday world, and until recently of the world of the physicist. In works of visual art the eye sees not simple spatial relationships of determined mutual positions of things. ''What the eye sees are tensions and countertensions, harmonies and disharmonies: purely dynamic relations . . .. The space of the picture itself, together with the things represented in it, is not simply set off from the observer; rather it opens itself to him, takes him into itself, passes into him.'' [20] It is this type of conception of space that has moved into physics and taken it away from the rigid structure of the space of classical geometry to a more flexible and dynamic expanding and contracting space as process.

He finally brings his argument together in a way that allows us to relate it quite explicitly to the temporal axis of intention. ''It is not the series of instant after instant which is essential in music, but the fact that the present instant contains the past instant and the future instant: an interpenetration rather than a succession. The same word by which we distinguished the order of auditory space is, then, applicable to the order of auditory time. Space and time: not 'juxtaposition' and 'succession,' but 'interpenetration'—interpenetration of simultaneous occurrence and serial occurrence.'' [21]

In effect, music and art—both of them the output and expression of creative human activity—cannot be considered simply as events or things describable in the four-dimensional world of three spatial axes plus the temporal axis of succession. They acquire their unique characteristics only when the temporal axis of intention, with its interpenetrating present, past, and future, its directionality, its force of goal-directed activity, are brought into the analysis.

Listening to a piece of music and looking at a work of art are intentional goal-bound human activities which come to life only when seen as human episodes, including episodes with interacting human beings. The active, temporally intact, and continuously unfolding event is what life is about. A musicologist has reminded us that this is as true for the spatial output of human action as it is true for the temporal outputs expressed in music. The human present is extensive: it has a plan; it has temporal and spatial order; it has living memory, living perception, and living desire and intent. Life is lived in timeless moments only in a conception of eternity. On earth it is lived in time-filled episodes, within a temporal frame which I shall consider in the next chapter.

[20] Ibid., p. 345.
[21] Ibid., p. 347–48.

CHAPTER EIGHT

The Time-Frame of the Individual

The two-dimensional conception of the temporal world in which human beings live has provided a construction which makes it possible to locate a unified field-of-force of memory, perception, desire, and intention along an axis of simultaneous past–present–future, and to date and record the particular times of successive changes in these mental states. This conception brings meaning into the analysis of behavior, through goal-directed episodes with planned procedures and targeted times for achievement of goals, in concert with the objective recording of the progress and actual duration of episodes. Such episodes include social interactions both spontaneous and institutionalized.

The question may be raised, however, whether this analysis is anything more than just an empty abstraction, just a meaningless mode of representing intention, or whether something more substantial can be achieved by its use. I believe there is substance here; and I propose to show in this chapter and the next why that is so. Let me set out my objectives in this regard as clearly as possible.

My main objective will be to illustrate by means of one particular set of examples the type of analysis of human behavior which becomes possible the moment a 5-dimensional world with a 2-dimensional time-frame is used as the starting point for observation and enquiry. I do so not with the notion that this particular set of examples is exhaustive, tells the whole story, is the end of the matter. On the contrary, I do so as a means of demonstrating by example a different type of approach to the understanding of psychological and social issues, an approach which may open up new vistas in the human sciences.

The particular examples I shall use are concerned with the measurement in time of tasks assigned in employment work. Such assignments can be measured objectively in at least two respects: first, they always begin with a targeted completion time, which may or may not be achieved, and which can be mapped on to the temporal axis of intention; and second, they always have an eventual completion time, so that the actual duration of the episode, the time taken for achievement of the goal, can be mapped on to the axis of succession and compared with the intended time.

My description will comprise two main parts, one factual, the other speculative. The first part will present a brief review of previously reported data built up over many years which suggest that the subtle feeling of the weight of responsibility in an employment role can be defined and objectively measured in terms of one factor alone, namely, the longest times targeted for the completion of assignments in that role. These targeted time-spans must be mapped on to the axis of intention, not succession. I shall refer to them as the *time-span of discretion;* and I shall contrast them with the actual time taken, which I shall call the *time-span of achievement*.

A range of practical consequences of the measurement of level of responsibility in terms of time-span of discretion will then be mentioned, including the possibility of achieving an equitable distribution of wage and salary differentials or relativities, and the discovery of certain general properties of bureaucratic hierarchies. It was, in fact, my seeking for some explanation for these findings, all steeped in the time structure of episodes, that was the original stimulus to my present interest in the general problem of the nature of time.

In the second part of this description I shall turn from the solid ground of findings to the less secure ground of speculation. These speculations will be concerned with an extrapolation from the idea on the the one hand that the felt weight of responsibility in a role can be measured by the maximum time-span of discretion in that role, to the idea on the other hand that a person's level of capability in work—the overall capability—can be measured in terms of the longest assigned time-span of discretion he is able actually to achieve, that is to say, of his maximum time-span of achievement. I shall refer to a person's time-frame in describing maximum time-span of achievement of which he is capable. My hypothesis then will be that the level or size of a person's capability in work can be defined by his time-frame. This time-frame is a reflection of the time-spans mapped onto the axis of succession for which he is capable of planning goal-directed episodes and of executing those plans and achieving the goals in the allotted time.

A person's time-frame—the time framework of the very longest time-spans of achievement which he can accomplish at any particular stage of his life—sets the total temporal domain within which the individual lives: it defines the limits of his active present. The conclusion may then be reached that different individuals have different time-frames and occupy temporal domains of different size. Hence, they may construct markedly different pictures or conceptions of the world both in scale and in quality.

It will be apparent that the whole of this second part of the argument—that concerned with the measurement of scale or level of individual capability in terms of size of time-frame—depends upon the possibility of discovering the longest goal-directed episodes an individual is capable of handling in any areas of his life—whether in employment, or in recreation, or hobbies, or other areas of personal interest. This problem of measuring maximum time-frame and size of temporal domain will therefore be explored. It will take us into some of the most fundamental questions of how people organize their purposes and intentions, set their goals and aims, and maintain the multiplicity of activities which con-fjstitute their active present; and of how, in so doing, they fix the boundaries of their own personal worlds and cooperate in the doings of their society.

Moreover, the question of measuring time-frame and temporal domain will call for a further sharpening of the concept of goals. In particular it will be necessary to distinguish between the subjective fluidity of private personal goals and the objective formulation of publicly stated goals set in cooperation with others or even by contractual commitment. These distinctions will take us into a number of aspects of social responsibility and social commitment.

Finally, there is one other set of findings which may help to establish an accurate measure of the temporal domain of individuals. They have to do with the possibility that temporal domain matures and unfolds in a regular way throughout life. If so, then to establish temporal domain at those points in his life and development when a person was heavily occupied in all-out application of his capability, would allow his own personal level and growth in level of capability and his future potential growth in level to be established. These questions of growth of time-frame and of temporal domain will, however, be left for the next chapter.

It may have been noted that time-span measurement is equal–ratio–length scale measurement. Some of the more general implications of this application of objective equal–ratio–length scale measurement to human affairs will be left to the final chapters. But in all these considerations the focus is upon the perspective to be gained by a proper temporal approach to the understanding of psychological, social, economic, and other human processes and events. In the form of time is to be found the form of living.

Time-Span of Discretion and Felt Weight of Responsibility

Central to my immediate argument is the general finding that in employent work—that is to say, work in bureaucratic systems in which people are gainfully employed for a wage or salary—the sense that each person has of the size of his job or, in other terms, his sense of the weight of responsibility of his job—varies directly and uniquely with the maximum periods of time he is required to exercise discretion in carrying out his assigned duties. I have termed this maximum period of time the time-span of discretion in that role.[1] Here is how the findings were obtained.

The essence of time-span measurement in employment roles lies in the fact that every assignment from a manager to a subordinate is made up of the following elements: first, a statement of what it is that is to be done—the required output or result, together with the prescribed limits of method, expense, routines, etc., which constitute the explicit conscious context within which the assignment must be carried out; second, a statement of the quality of the output—the limits of quality governing the subordinate's exercise of discretion; and third, the maximum time the manager decides (for whatever reasons) to allow for the completion of the task—the maximum target completion time. This combination of an output, within a prescribed context and given quality limits, by a given maximum target completion time, defines the goal for the subordinate.[2]

It will be apparent, however, that the goal and task assigned by the manager define and set out in a very precise way the manager's intention. It is this intention which the subordinate is employed to take over and accept as his own intention, and contractually undertakes to carry out. That contractual obligation is in the nature of employment work. The transfer of intention from a manager to a subordinate constitutes the psychological core of manager-subordinate relationships.[3] I shall argue in the final chapters that these relationships are a prime example of the way in which investment in common intentions and the common testing of outcomes are major elements in how one gets to know the mind of another.

The manager's goals or intentions, which he transfers to the sub-

[1] The findings about time-span of discretion and its relation to felt weight of responsibility have been reported in the following series of books. The first report of time-span was in Jaques, (1956), *Measurement of Responsibility,* followed by *Equitable Payment* (1961) and *A General Theory of Bureaucracy* (1976) by the same author. The most systematic study of time-span of discretion as related to equitable payment (a relationship to be described later in this chapter) was carried out at Honeywell Corporation in Minneapolis, U.S.A., by Dr. Roy Richardson and reported in his *Fair Pay and Work* (1971).

[2] See Jaques, (1964), *Time-Span Handbook,* for a detailed description of the method of time-span measurement, with many examples.

[3] I shall not elaborate on this theme here, but have done so in *A General Theory of Bureaucracy,* Chapter 4.

ordinate, are locatable on the temporal axis of intention. The vicissitudes of the goal, and progress toward the goal, can be tracked on a series of axis of intention cuts. The actual length of time of achievement of the goal is measured retrospectively along the temporal axis of succession—the length of time from beginning of task to achievement of goal.

The maximum spans of assigned goals in an employment role give the measure of the time-span of discretion in that rôle. These time-spans can be objectively ascertained by talking with the manager and assisting him to explicate his decisions with respect to targeted (intended) completion times for his subordinate's assignments.[4] The peculiar significance of the time-span of discretion does not lie simply in the fact that it can be objectively measured. It lies in a number of related findings.

The first of these findings is that as the time-span in a role is either increased or decreased through changes in work circumstances, the weight of responsibility in the role is experienced by the incumbent of the role as increasing or decreasing. Thus, for example, if a manager wishes to decrease the weight of responsibility in a role in order to accommodate a new and inexperienced subordinate who is in training, he will nearly always divide the longer-term assignments into two or three parts and have the subordinate complete each part in turn; the effect is to decrease the time-span. Equally, as the subordinate becomes more experienced, the manager will assign the longer-term tasks as a whole, thereby increasing the subordinate's responsibility and, concomitantly, the time-span of discretion in the role.

But this fluctuation of felt weight of responsibility with time-span does not necessarily prove that a given time-span is uniquely related to a given level of work. The possibility of such a unique one-to-one relationship emerged from the dramatic finding—which has been repeatedly obtained—that the time-span of discretion in a role correlates almost perfectly with the incumbent's iudgment of what constitutes fair pay for the work he is being given to do regardless of any other factors. These correlations in various studies range between 0.85 and 0.92.[5]

Here then is some very considerable evidence of the significance of the organization of human intentionality in time. Regardless of whether

[4] I shall argue in Chapter 11 that time-span measurement is a first example of the development of an objective equal–ratio–length scale giving a derived measurement of meaningful or purposeful human activity, and that it demonstrates the value of adopting a 5-dimensional framework.

[5] Detailed reports of these findings (including a critical evaluation of some apparently negative findings), obtained in over twenty different countries, will be found in Jaques, *Equitable Payment,* and in Jaques, *A General Theory of Bureaucracy,* (Chapter 14 contains a critical review of findings). Systematic controlled studies are reported in Richardson, *Fair Pay and Work,* and in a comparative study of 12 British firms reported in George Krimpas, (1975), *Labour Input and the Theory of the Labour Market.*

a person is employed as a nurse, manual worker, manager, salesman, production controller, planner, personnel officer, research technologist, computer programmer, production engineer, chemist, regardless of the apparent or assumed level of that employment, and regardless of the actual wages and salaries being paid, people working at the same time-span of discretion name the same level of pay as fair and just for their work. I have been able to think of no other sound explanation for this phenomenon than the fact that the experience of level of responsibility, the sense we have of the weight of a particular role, is uniquely related to the maximum time-span of discretion we are required to exercise in that role.

I would emphasize that I am referring to the maximum time-span of discretion of all tasks in a role, and not of any one task by itself. It is the role time-span which gives the measure of the overall planning and foresight that is required, not just for any one task but for the organization and carrying forward of the whole complex of tasks in which a person might be engaged at any one time in his active present in the role. Any two roles which require the incumbents to work the same distance forward in time as mapped onto the axis of intention feel as though they have the same level of work.

This hypothesis about the direct relationship between the time-span of discretion and the felt level of work was given what I have taken to be almost incontrovertible support by the subsequent discovery (in 1957) of an underlying systematic structure of managerial levels of organization of bureaucratic hierarchies. This underlying structure shows discontinuities and the emergence of successive true managerial levels at measured time-spans of three months, one year, two years, five years, ten years, and higher. In short, a role at one-year time-span of intention will occur at the same real level of bureaucratic organization regardless of type of institution—industrial, commercial, civil service, health or social service—in the United Kingdom, France, the United States, Canada, Sweden, Australia, or in any of over twenty other countries. The time-span of discretion appears to be a measure of level of work in organizational terms.[6]

Time-Span of Achievement and Level of Capability

The foregoing findings have led to the formulation of proposals for sweeping changes in our methods of handling one of the most vexing questions of industrial societies, that of how to establish wage and salary differ-

[6] I have described these findings in *A General Theory of Bureaucracy*.

entials which will be experienced as fair and just.[7] They have also been used in practice for constructing in a systematic way managerial organization hierarchies which have what feel to be a comfortable or right number of levels—neither too many nor too few.[8]

These practical consequences aside, however, the question which faces us is what explanation can there be for this seemingly curious significance of time-span of discretion for the individual. It is at this point that I must leave the realm of objective findings and turn to a series of speculations and hypotheses—speculations which are, however, supported to some extent by findings.

The proposition I shall pursue is that the time-span of discretion gives a measure of felt weight of responsibility in work because individual capability to do work and carry responsibility is organized in time, is expressed in time; and, moreover, that the level of an individual's capability in work is directly measurable in terms of his time-frame—the longer the time-frame which a person can plan and with which he can cope in practice, the higher his level of capability. In other words, the secret of what constitutes the level of capability of a person in work lies locked up in the two-dimensionality of time in the living world—locked up, that is, in the relationship between, on the one hand, intention, goal-setting, and planning mapped onto the axis of intention, and, on the other, the expression of plans in action and actual achievement of goals mapped onto the axis of succession.

It is difficult to describe the phenomenon which I am trying to identify, and which I am referring to as level of capability as measured in time-span.[9] A person's time-frame is not how long into the future he happens to think or fantasy, or in the case of a child, how long it can aimlessly dawdle. It is concerned with specific, identifiable activities, with specific goals, with specific completion-time targets (even though those completion times might be changed during the course of the activity). It is what a person is able actually to plan for and to achieve and not just to wish for or to happen upon by accident. Time-frame gives a

[7] Wilfred Brown, (1973), *The Earnings Conflict;* Jaques, *A General Theory of Bureaucracy;* John S. Evans, (1979), *The Management of Human Capacity*.

[8] Jaques, *A General Theory of Bureaucracy;* Wilfred Brown, (1971), *Organization;* Social Services Organisation Research Unit, Brunel University, (1974), *Social Services Departments: Developing Patterns of Work and Organisation;* Jaques, (Ed.), (1979), *Health Services*.

[9] I have previously used the term "work capacity" for this feature of behavior, in *A General Theory of Bureaucracy*. Capability has been usefully defined by Gillian Stamp as the overall ability to pattern and order experience in space–time. ("Assessment of Individual Capacity" in Jaques, Gibson and Isaac, (1978), *Levels of Abstraction in Logic and Human Action*.). Muller and Van Lennep have also connected time-scale with level of work in their formulation of their "helicopter principle"—that is to say, the higher the perspective an individual can exercise, the higher his capacity will be, and the farther into the future he will be able to plan and act. D.J. van Lennep, (1968), "The Forgotten Time in Applied Psychology."

measure of the level of that overall capability of an individual which enables him to carry higher or lower levels of responsibility in work, in actually doing things. It is what selectors look for when they ask, "Is such-and-such a candidate big enough for the job?"

Level of capability is what psychologists have hoped to measure by intelligence tests such as Raven, Binet, Wechsler, and Rorshach (on ability). The language and concepts of intelligence and of IQ, however, are unfortunately of no help, for intelligence expressed in IQ does not correlate very significantly with the differential ability among people to carry very different levels of work in their occupations. Low IQs and low intelligence test scores are likely to be inconsistent with carrying high levels of responsibility; but high scores do not necessarily guarantee the ability to do so. I now believe that the shortcoming with these intelligence testing procedures is that they are 4-dimensional in approach. To the extent that they take time into account it is by limiting given tests to a particular amount of time (axis of succession only). They do not incorporate any conception whatever of measurement along the axis of intention: intentionality, and achievement of intention, those most important features of human capability, are simply left as floating variables.

A person's time-span of achievement, in contrast to 4-dimensional IQ measurement, gives an indication in quantifiable terms of his ability to organize in an active and meaningful way his current world. It is a 5-dimensional conception in that its measurement calls for finding out both about the person's intentions and goals (ti-axis) and about his success or failure in carrying out those intentions (ts-axis). Or in other terms, it indicates his ability to organize his total active working relationship with his world, a working relationship in which he is actively engaged in striving to transform his world—both external and internal—in accord with his needs and in order to satisfy those needs in a planned, organized, and intentional manner.

It is this work activity subsumed under the concept of capability which the data about time-span and felt-fair pay, and time-span and levels of bureaucratic organization, suggest is directly measurable by time-span of intention and achievement. The argument is a simple one. If, in order to carry higher levels of responsibility in employment work, an individual must work at longer time-spans of intention, then his time-frame of actual achieved times must be greater. Level of work and the kind of capability that must accompany it are both encompassable within a 2-dimensional time framework.

Temporal Horizon and Temporal Domain

The findings I have just described have a major shortcoming from the point of view of ascertaining an individual's maximum time-span of achieved intentions. They are derived solely from studies of individuals employed in bureaucratic systems. It may be the case, however, and indeed often is, that a person's full capabilities are not exercised in his employment. The person may not seek to do so, or perhaps may not have the opportunity to do so. In order to generalize my formulation, I wish now to establish two concepts which apply not just to time-span in the employment situation but to time-span as it occurs in a person's life in general, in his active present as a whole, in his home life, in recreation, in social and political activities. These two concepts are temporal horizon and temporal domain.

I shall borrow a formulation from Paul Fraisse,[10] and shall define the *temporal horizon* of an individual as that goal among all those toward which he is working at any given time which has the longest forward targeted completion time. It is the longest forward planned task in his active present, the farthest forward that he is looking at that moment. The distance of the temporal horizon thus changes as goal-directed episodes are completed and replaced by others, or targeted completion times of existing episodes are changed; just as the distance of the geographical horizon changes with changes in vantage point, or terrain, or visibility.

But just as the geographical horizon has a maximum distance depending on the height of any given vantage point, so also, I would hypothesize, will any given person's temporal horizon have a maximum distance depending upon the level of capability of that person. The level of capability of the person (his maximum time-span of intentional achievement) sets a person's vantage point with respect to the temporal horizon, just as height above sea level sets the vantage point for the physical horizon. I shall refer to the maximum temporal horizon of a person as his time-frame, and to the temporal area bounded by a person's time-frame as his *temporal domain*. My assumption is that a person's time-frame and related temporal domain give a direct measure of his level of capability.

Each person's field of action extends continuously to the boundaries of the temporal domain bounded by his time-frame encompassing all shorter episodes, in the same way as his field of perception extends

[10] Paul Fraisse, (1964), *The Psychology of Time,* Chapter 6, pp. 151 to 198. He uses the phrase "temporal horizon" to refer to the way "each of our actions takes place in a temporal perspective." He does not state whether this temporal perspective refers to a multiplicity of tasks, nor does he state where it comes from. But his general idea does seem similar in many respects to my own.

continuously to the geographical horizon and can take in everything between. It is the longest distance along the ti-axis (the temporal axis of intention) which he is capable of successfully spanning or bridging by trajectories of purpose and action. Beyond the end of these bridged activities, beyond the farthest limit of the temporal domain that is, lies an increasingly fragmentary and discontinuous mental construction of the past and future.

The concept of temporal domain is concerned with the world of action and not with the world of imagination and fantasy. The historian can, for example, reach way back into antiquity and beyond. But he makes his way there by great mental leaps across intervening periods in which he is not interested. He does not encounter the distant past across a continuous and illuminated field of activity. His actions, and thus his temporal domain, are bounded by the time-scale within which he plans and carries out his historical study and research programs, and not by the dates of the epochs which he chooses for study.

Similarly, we may reach in our minds into anticipations and hopes for the future across lifelong leaps even into old age. But we cannot (unless we inhabit a huge temporal domain) detail and plan a continuous and unbroken trajectory between the "now" and "then" on the way. Beyond the boundary of the temporal domain, isolated planned or desired events lie dotted about in the anticipated future on the ti-axis to be picked up and acted upon at some appropriate time—until our envisaged death. Beyond our own death there lies the imagined future of other people. But as far as we ourselves are concerned there lies either nothingness or eternity, depending upon our own religious or philosophical outlook.

Man as a Multiple-Episode Creature

The temporal domain thus encompasses all the current goal-directed episodes in which a person is engaged. For one of the outstanding characteristics of people is that they are always involved in more than one goal-directed course of action at any one time—we are multi-action creatures. Each person is engaged in many activities, in the pursuit of many goals at the same time—one activity being forwarded a little, then another, then the first again, then another, as opportunity or desire arises. Some activities are short and may be completed at one go, such as, for example, eating a meal alone, or seeing a film, or having a particular and limited conversation with someone, say an interview. Other activities are extended in time and pursued intermittently, such as, for example, studying for exams, buying a house, or starting a club. And even many apparently continuous transactions are in fact intermittent when examined more closely. Thus, for example, a business lunch may be composed of two intermittent transactions—having a good meal, on the one hand, and forwarding the business conversation, on the other.

This multiplicity of purposive trajectories is an important feature of human behavior.[11] It requires judgment. It requires a sense of priorities. Some people are good at it. Some people are extremely bad—putting off what are really their most important activities while pursuing zealously—even compulsively—the less important ones. Or they may put aside the more difficult activities, or less desirable ones, or the ones that for other reasons may seem most burdensome and unattractive, trying to stave off the inevitable day of reckoning when what ought to have been done emerges as not yet done, and the consequences of the incompleted task must be faced.

When I say that everyone is moving along a multiplicity of trajectories, I would exclude the more seriously mentally handicapped individuals, who may become confused if they are required to cope with more than one thing at a time. This shortcoming also shows up in certain types of brain-damaged patient.[12] But in normal circumstances multiplicity of purposive activity is the ordinary everyday state of affairs for everyone.

This patterning of multiple activities can be illustrated in the assemblage of goal-directed episodes depicted in diagram 7.3. The following points may be noted. First, the goals for the particular person range from an hour or so distant in time to a distance of many months. Second, the longer-term episodes overlap one another, and overlap the shorter-term episodes, so that many episodes are running concurrently at any given moment. These concurrent trajectories oscillate between figure and ground, with only one the subject of consciously focused activity at any time.

Third—and this feature is the one which gives the definition of size of time-frame and temporal domain—*the longest oscillatory interweaving of goal-directed trajectories is determined by the longest of the planned trajectories in the assemblage*. This feature can be illustrated in the diagram: on March 1, for example, the limit of the person's temporal horizon was $5^1/_2$ months; and in August the limit of the temporal horizon was nearly 5 months. Hence his time-frame is at least $5^1/_2$ months, but might of course be longer. At each of these times the person had alternately to push forward the longest planned tasks ($5^1/_2$ months and 5

[11] The notion of multiplicity of trajectory would fit the multiple-drive theory of motivation as against the single-drive theory. Stephen Pepper has shown the philosophical significance of multiple-drive theory. It gives a structure combined with flexibility to human values and ethical choices. Pepper, (1958), *The Sources of Value*, pp. 143 to 150.

[12] As described in K. Goldstein and G. Scheerer, (1939), *The Organism*.

months respectively), (or to decide to leave them cooling in the background), as well as to forward a number of shorter tasks, and to carry out during the day a whole series of tasks lasting minutes and hours which make up the ordinary round of activities of everyday life.

Handling a multiplicity of simultaneously running tasks is a complex business. It is like taking care of a crowd of restless children in a playground, all of whom have to be watched out of the corner of your eye while you deal with one most urgently needing attention; or perhaps like the entertainer who sets a large number of plates rotating on bamboo poles and has to keep running from one to the other to give the poles a shake to keep the plates spinning.

The reason why the task with the longest planned or targeted goal-directed trajectory sets the limits of the time-frame of the individual may be self-evident. It is that trajectory which sets the limits within which all the others are contained; that is to say, it comprehends all the others in having to be kept in the back of one's mind and brought into focus from time to time, while all the shorter tasks are catered for and with countless very short tasks being started, carried through, and completed while the longer ones are attended to as necessary.

It is the goal-directed activities of people which define their active life.[13] All concurrent goal-directed activities are inevitably interactive, since they compete for the person's focus of attention, compete to shift from ground to figure. The longest in time of these tasks sets the outside limit to the actively engaged mental organization, including both the changing figure and the complex field of the ground.

Private Goals and Publicly Contracted Goals

If temporal domain is to be defined in terms of the maximum length of purposeful episode which an individual can achieve—his maximum time-span of achievement—how is this maximum time-span to be discovered? One thing is clear, and that is that the whole of a person's active life must be explored, and not just his employment, in order to be sure of taking into account all his purposeful activities including those which he has more freedom to set up for himself than he might have in his employment.

Take, for example, the case of a father and businessman who is

[13] Susanne Langer has placed great emphasis upon the act as a central feature of life. "It is with the concept of the act that I am approaching living form in nature, only to find it exemplified there at all levels of simplicity or complexity, in concatenations and in hierarchies, presenting many aspects and relationships that permit analysis and construction and special investigation. The act concept is a fecund and elastic concept." Susanne K. Langer, (1967), *Mind: An Essay on Human Feeling*, p. 261.

employed in a research department where he is assigned a program with a multiplicity of research tasks, some to be completed in days, some in weeks, some in months, and some extended programs to be progressed intermittently, with delegation to subordinates, and targeted for completion in, say, a year and a half; at the same time he has to plan and carry through the redesign and reconstructing of his garden during nine months; do some special home study on his own for a special qualifying examination in a year's time; carry out day-to-day tasks at home, such as minor repairs, washing the car, cutting the grass, as well as moderately longer tasks, such as spending a month in getting the family's camping gear together, repaired, and in good order for the summer vacation.

Or, to take another example, consider the purposive activity of a mother and house-wife, who must plan meals for the next few days; prepare a special dress for her daughter for a party set for the weekend; fit in certain sewing and mending and washing during the week; plan and prepare a program for a Women's Institute bazaar in the coming month; arrange time to help her son with his review studies on a regular basis for examinations coming up in two months' time; plan and carry through, with some help from her husband, the redecoration of three rooms, to be completed over a three-month period; plus myriad other day-to-day and week-to-week tasks arising unexpectedly, like some plumbing repair work, an unexpected guest to be fitted in over the weekend, and so on.

These two examples may be readily extended to cover everyone. In some cases the longest term of the goal-directed episodes may be connected with goals perhaps a maximum of a week or so forward—life being lived on what would feel more like a day-to-day basis. At the other extreme will be found families in which the father or mother or both have some goal-directed episodes some years or many years into the future. But in practice it can prove to be much more difficult than these examples would indicate to establish the planned completion time for personally set tasks. In the employment situation, this problem does not arise, since the target completion time is fixed as an objective fact by a person's supervisor and can be changed only if objectively changed by the supervisor.

In private life, however, it is not necessarily always clear what a person means when he says to himself that he must "fix such-and-such in the next few months," or "learn to speak French this year," or "get his golf handicap down to 5 within three years." Does he really mean that he will pursue these activities continually as part of his active present? Or does he mean that if circumstances work out he will try to do these things, and if not, that he will let them slip or put them off until later, or perhaps not bother with them at all? Or does he mean, even, that he thinks in a general way it would be a good idea if he were to pursue these activities but he has not really committed himself to doing so? It is impossible even for the individual himself to know quite what he means, for good intentions, like New Year's resolutions, are not always very real and can vanish as readily as they appeared.

It is in fact not possible to establish the achieved time-spans for private personal goals. Such goals are inevitably fluid. Because they are not shared, they are not objectified and cannot be fixed and gripped. They are held only to the extent of the will and self-discipline of the individual. Therefore, when a goal is achieved, if it is, it cannot be said that it was completed as intended and planned, because the plan itself was inevitably floating, tossed about by circumstance.

What can be done, however, is to focus the analysis and measurement upon objectively established goals; that is to say, upon goals which have been fixed in relation to someone else, agreed with the other, and related to that other in the sense that the activity is carried out on his behalf. All employment work has this feature. But it is possible to contract to do things for others under many conditions: to provide a service, to knit a sweater in exchange for something else, to give a series of lessons, and so on. Once a goal has been established in this way it can be said objectively to exist; someone else will judge whether it has been achieved in the agreed way by the agreed time; and if there is disagreement about whether or not the bargain has been kept, a third party can be called in to judge.

This approach to the measurement of achieved time-spans, and thence to the establishment of time-frame and temporal domain, in terms of objective socially agreed and contractual goals appears at first to impose an unfortunate limitation upon our analysis. Further reflection, however, suggests that it might be wrong to jump too quickly to any such conclusion. For time-frame and temporal domain in terms of objectively defined achieved time-spans would give us an assessment of the competence of the individual to carry through purposeful behavior in relation to others in a socially connected way. It may not tell us what he might do in his own private world; but it would tell us about his social competence, about his competence to function in a setting of social responsibility and committal, and respect for the needs of others.

The exigencies of practicality have, therefore, pitched us into a social definition of a person's time-frame and of the scale of his temporal domain and capability measured in size of time-frame. It is the size of the domain in which he lives and is gripped in interaction with others, each occupying his own temporal domain. The rootless person, the aimless person, the true bohemian, whatever the reason for his social alienation, would not have a measurable temporal domain, whatever might be his fantasies and his private intentions.

In short, we may redefine a person's temporal domain as that complex of goal-directed episodes which have been objectified by his having committed himself to others for the achievement of the goal and by an agreed time, its outer boundary being set by the longest distance forward of planned episode which he is capable of achieving to the agreed and contractually bound standard.

Goals within the Active Present and beyond It

There is one final problem associated with the ascertainment of a person's time-frame and temporal domain which I must dispose of before turning, in the next chapter, to the question of whether a person's temporal domain is a genuine quality of a person and, if it is, how it grows and develops throughout life. This problem is that of goal-directed behavior which seems to be lifelong in extent. There are undertakings, for example, such as the payment of insurance policies, or the purchase of a house by mortgage, or long-term saving for a child's university education, or perhaps a five-year-old's stated intention to become a policemen when he grows up, which seemingly have enormously long time-spans of intention, and which, if carried through, seemingly have equally long time-spans of achievement. Moreover, since most people have insurance policies or mortgages, and most children have ambitions to be something or other when they grow up, it would appear that, contrary to my argument, everyone has roughly the same size of temporal domain and the same level of capability.

There is a fallacy in all these instances, however. It lies in the fact that once the selection of the insurance policy or other arrangement has been completed and the mode of payment arranged, no further decisions *with respect to the eventual goal* need be taken. Periodic payments are fixed, and getting together the necessary sum for each payment constitutes a discrete goal which fits into the priorities for that week, or month, or other period during which the active getting together of each payment occurs. But no further work will need to be done with respect to the policy itself. That is to say, getting to a fully-paid-up pension by the age of 65 is achieved by buying a prearranged policy and payment schedule at the age, say, of 30 years—the task of purchasing the policy requiring say a few weeks or months to complete—followed by a succession of discrete monthly tasks concerned with accumulating the funds for each payment. It is these discrete goals and tasks only which enter into competition with other episodes for priority focus, and which therefore determine the temporal horizon and fall within the temporal domain. The eventual outcomes are part of the long-distance unknown beyond the boundaries of the person's temporal domain.

Another type of activity in which the longer-term aspects are planned and arranged by others is exemplified by the creation of a garden where a plan is purchased which describes each set of activities month by month, the gardener being required to organize only the month-by-month goal-directed episodes; or the making of an extensive wardrobe, where the person buys the patterns and the worked-out planning and sequence, and simply follows it step by step. Such purchased plans—a kit for making a boat is another example—make it possible to carry out the activity on a longer or shorter basis, depending on how a person breaks down the various parts and builds them into his or her schedule of other actively present activities.

A goal-directed episode, then, is one in which the movement toward the goal is continuously alive; the person is continuously on the go; the goal is continuously in mind and intermittently in focus. It is "continuously" rather than "continually" in mind in the sense that active conscious preoccupation may be intermittent, but underneath, in the unconscious behavioral field, the activity is continuous; it is always under pressure; there is a constant predisposition to the goal, and the issue simply forces itself into conscious attention from time to time. The path to the goal is being continuously traversed, and when necessary modified; the activity appears in dreams; solutions to difficulties occur during sleep and appear ready-made on waking—indeed, this is how creativity occurs during sleep.

In other words, a goal-directed episode is a vector—it is a psychic system under continuous tension—the tension not being relieved until the goal has been reached or abandoned, and an end-state feeling achieved whether of satisfaction or dissatisfaction. It is a multiplicity of these vectors, these active goal-directed behaviors, these goal-directed strivings, which make up the everyday mental life of human beings. And each one has to select his priorities, plan his time—consciously or unconsciously—or some of the goals are not going to be reached, and frustration and dissatisfaction will result.

CHAPTER NINE

Patterns of Growth of Time-Frame

The general analysis of the nature of time and of the five (3 + 2)-dimensional structure of the social world has revealed the significance of separating out from each other the temporal axis of intention (ti) and the temporal axis of succession (ts). These axes have been used to establish the concepts of the time-frame and the temporal domain of individuals. It has been suggested further that the scale of a person's time-frame may give a direct measure of the level or scale of his capability. Level of capability has been defined in terms of the person's capacity to carry responsibility for higher or lower levels of work, in action, in doing things, in real life—it is a measure of the scale of a person's overall ability in action.

With the concepts of time-frame and temporal domain, we bring the problem of the nature of time into direct conjunction with the problem of the nature of human existence—socially and psychologically. I propose, furthermore, to link the growth of the conception of time, from childhood to adulthood, to the growth of a person's time-frame; one's sense of the meaning of time develops with one's growing time-frame and the accompanying foresight. Understanding time and doing things in time are parts of the same process.

Some of the difficulties which still have to be overcome in getting accurate measurements of time-frame have been discussed. The hypothesis has been suggested, however, that time-frame grows in a regular manner with age. If this hypothesis is correct, then the discovery of a person's time-frame at one or two points at any stage in his development would give a prediction of his potential, that is to say, of his likely time-frame and level of capability at subsequent stages in his development.

The hypothesis about the pattern of growth of time-frame is presented in this chapter, and its implications for the assessment of time-frame in children and in adults are described. It completes the illustration of my most general theme, namely, if we start with episode (temporal abstraction) rather than with point-at-a-distance cross-section (spatial abstraction), and with a clear understanding of the two-dimensional nature of time which is critical for the understanding of social and psychological phenomena, then human behavior can become the subject of scientific study without relinquishing its purposefulness or losing its humanity.

Time Perspective in Individuals

The assumption that the larger the time-frame within which a person can function, the greater is his level of capability, would be supported if it could be shown that people's time-frame increased in a regular way as they grew older. For such a growth would at least coincide with the fact that the scale or level of the capabilities of individuals increases with age—their competence in work, that is—and that this level of capability, unlike IQ for example, increases by and large throughout life.

Many psychologists have noted just this kind of growth. Fraisse, for example, has written: "The temporal horizon develops slowly through childhood. Its beginnings therefore present us with an excellent opportunity to start this analysis by establishing its *nature*. . . . at two months, for instance, a baby will turn his head in the direction from which he has heard a noise.[1] This utilisation of signals implies the existence of a temporal horizon, but at this age it develops entirely on the plane of what is being lived; the past and the future are simultaneously in his behaviour of the moment. Thus reaction chains are gradually formed in which each event becomes the signal for the next one. The way in which a child helps his mother dress him from the age of about ten months shows that he is becoming capable of adaptation to more complex temporal series. Later he will even take the initiative by performing the first gestures of a series, guided by more long-range anticipation; thus he will go and find his coat and shoes so that he can be dressed to go out.[2]

"In all these initial instances of temporally organised behaviour it is the past which gives meaning to the stimulus, turning it into a signal, but the signal gives rise to behaviour which is oriented toward the future."[3]

This finding is reinforced by experimental work with children, described by Fraisse, in which "A desired object is hidden before their eyes in a multiple choice apparatus and they are prevented from going to find

[1] Jean Piaget, (1937), *La Construction du Réel chez l'Enfant*, p. 326.
[2] Malrieu regards this behavior as make-believe, but we think it is of the same nature as that of the baby who stretches his arms out toward his mother to be picked up. P. H. Malrieu, (1953), *Les Origines de la Conscience du temps*.
[3] Paul Fraisse, (1964), *The Psychology of Time*, pp. 153–154.

it until a certain interval has elapsed. The interval may be increased according to age without impairing the success of the results. Obviously these intervals always depend to some extent on the situation but the important fact is that they increase with the age of the subject, whatever type of problem he is presented with. Hunter[4] used a multiple choice apparatus with three alternatives and found that the tolerated interval increased from 50 seconds at 2 years, 6 months to 35 minutes at 6 years. In connection with another problem, Skalet[5] found that an interval of a few hours at the age of 2 reached 34 days at 5 years, 6 months. As the child grows he becomes capable of taking into account in his activity the things which have preceded and which will follow.''[6]

Piaget also was well aware of the extension of the child's time sense with age,[7] and Fraisse summarizes the extensive writings on this theme by Stern, Decroly and Degand, Oakden and Sturt, Bradley, Ames, Gesell and Ilg, and Malrieu.[8] Malrieu in particular emphasizes this development, showing how the child increasingly organizes his memories (past experiences) and becomes interested in the past experiences of his parents and others.[9]

In the research of Lewin and his coworkers there are occasional references to time scale that are of some interest. They showed, for example, the reduction in time scale of young children under conditions of frustration.[10] In another article Lewin wrote: ''The infant lives essentially in the present. His goals are immediate goals; when he is distracted, he 'forgets' quickly. As he grows older, more and more of his past and future affect his mood and action.''[11] In this same article he goes on to add that the adult finally takes on very long time perspectives when he begins, for example, to think in terms of purchasing life insurance. As I have indicated, however, such activities do not necessarily point to a long time-frame. Lewin here lost sight of the critical issue of the individual's active goal-directed present, and therefore was unable to note the importance of individual differences in time-frame, to which subject I shall now return.

[4] W. S. Hunter, (1913), ''Delayed Reactions in Animals and Children.''
[5] M. Skalet, (1930–31), ''The Significance of Delayed Reactions in Young Children.''
[6] Fraisse, op. cit., p. 178.
[7] J. Piaget, (1946), *Le Développement de la Notion de Temps chez l'Enfant.*
[8] Fraisse, op. cit., p. 179
[9] P. Malrieu, (1953), op. cit., pp. 85–87.
[10] R. Barker, T. Dembo, and K. Lewin, (1941), ''Frustration and Regression.''
[11] Kurt Lewin, (1942), ''Time Perspective and Morale,'' p. 98. Quoting Lewin, Stephen Pepper states the view that, ''A man's life-space at any moment contains a time dimension. . . . With experience and maturity it tends to get longer. . . . A man's life-space tends to become less and less clear as it extends into the future. Nevertheless the more skillful a [man] is the farther ahead he looks and the more detailed are his anticipations.'' *The Sources of Value,* pp. 443 and 444.

Growth of Time-Frame: A Set of Hypotheses

In considering how best to convey a number of ideas in the socially sensitive area of how each person's time-frame and temporal domain (and possibly level of individual capability) might grow and develop with age, I decided that the best thing to do would be to begin by boldly setting out my hypotheses, and this I shall do. Then I shall explain where the hypotheses came from, and discuss some of the implications and consequences.

The array of curves shown in Diagram 9.1 sets out my hypotheses about the growth of time-frame in individuals from childhood to old age. Any given individual's maturation and growth in time-frame will follow one of these curves. Thus, for example, if at five years of age the scale of a child's time-frame is one hour, then the hypothesis would predict that his time-frame will mature and increase in scale to 7 hours at the age of ten, one month at the age of twenty, one year by the age of 55, and slightly above one year in old age.

Stated in more general terms, the hypothesis is that there is a regular predictable growth in scale of time-frame in ordinary life circumstances. Individuals at the same time-frame at the same age will follow the same pattern of maturation and growth. How far this pattern might be modified by special circumstances of social deprivation or of excess, and what such circumstances might be, I shall consider below.

The Time-Frame Progression Array

The curves describing the hypothesis of growth in time-frame have been derived from the following findings. These findings have been reported in detail in several publications.[12] I shall briefly summarize some of the main points, starting with data about earning progressions. I shall then show how these earning progressions can be translated into a hypothesis about progressions in time-frame.

The first finding was that there is a strong tendency for the total wage and salary compensation of employed persons to follow a regular pattern when corrected to a common value by means of the national index of earnings so as to eliminate the effect of general increases which apply to everyone. The rate of relative progress is by and large the same for everyone of the same age and earning level. The same general trend was found to hold for all types of employment in over twenty different countries. The general pattern of progression is as shown in Diagram 9.2. In this diagram each curve is an earning progression for one person, corrected

[12] Elliott Jaques, *Equitable Payment; Progression Handbook;* and *A General Theory of Bureaucracy*.

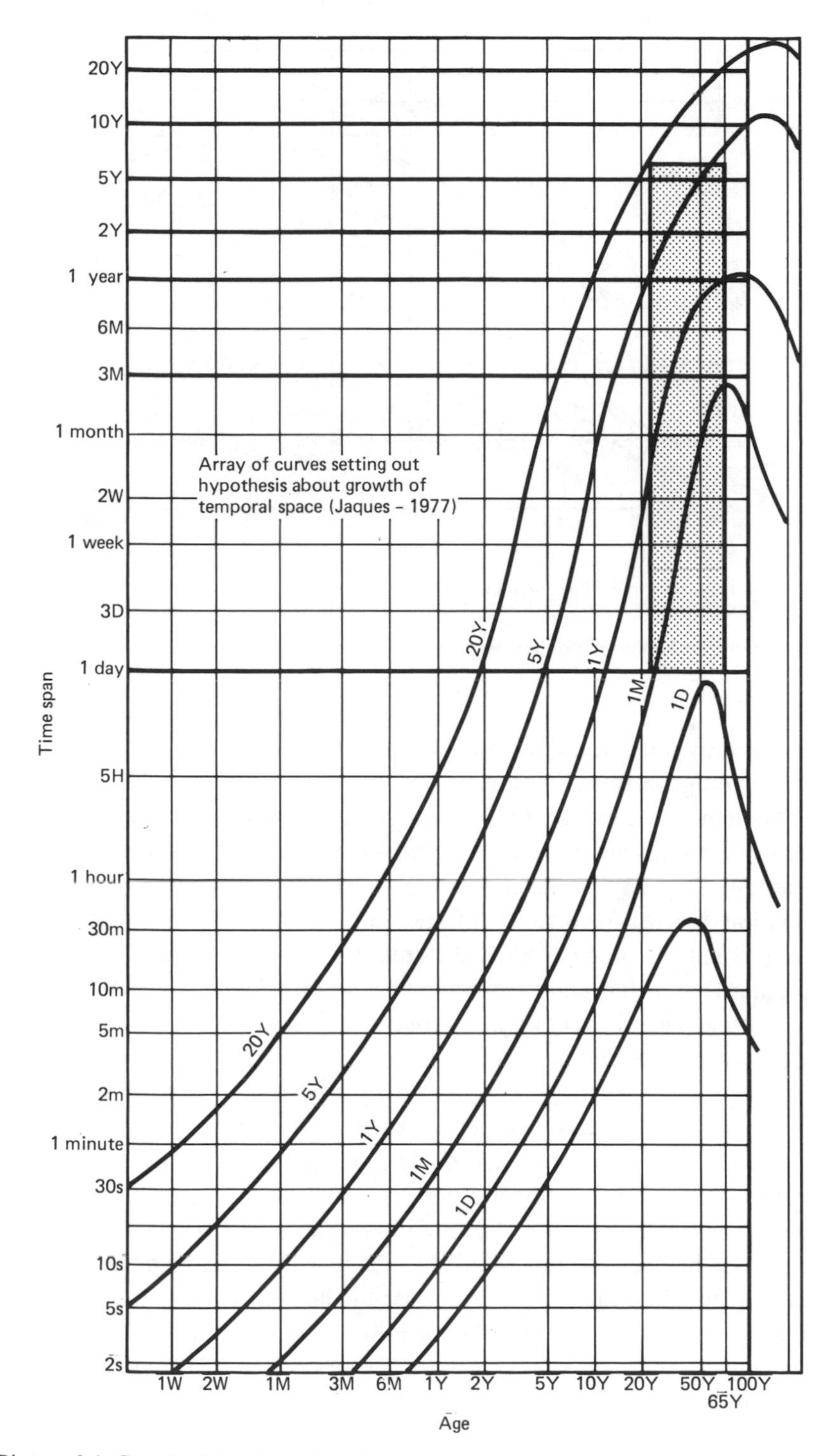

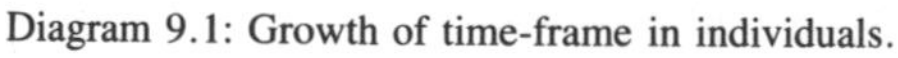
Diagram 9.1: Growth of time-frame in individuals.

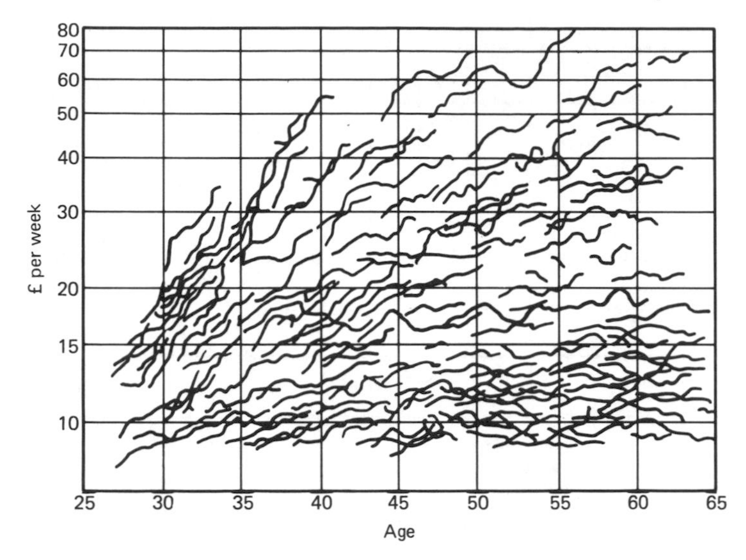

Diagram 9.2: Pattern of Individual Earning Professions.

so that all movements, up or down, are movements *relative to* the movement of average earnings in the nation.

The general trend of these curves—and note that they represent movements in individual earnings *through time*—was abstracted as shown in Diagram 9.3. My preliminary hypothesis in so doing was that the explanation for the recurrence of the same general trend in so many different countries, under widely different economic conditions, was that the regularities in the pattern were an expression of the rate of growth of level of individual capability employed in work and of the concomitant rate of growth in economic reward.[13]

This preliminary hypothesis about growth of capability was then tested against individuals' expectations of growth in earnings. Here a very striking finding was obtained. The progression in earnings which individuals sought as a right for themselves followed closely one or other of the curves in Diagram 9.3. Let me illustrate. If you ask any employed person what he feels would be a fair and just total compensation for a job that he would consider just right for his current capabilities, he will name a figure—say X_1. If you then ask him to think of a job that would be just right for him in, say, five years' time, and to consider what total emolument (under constant economic conditions) he would anticipate as

[13] This hypothesis was first set out in *Equitable Payment,* Chapters 9 and 10.

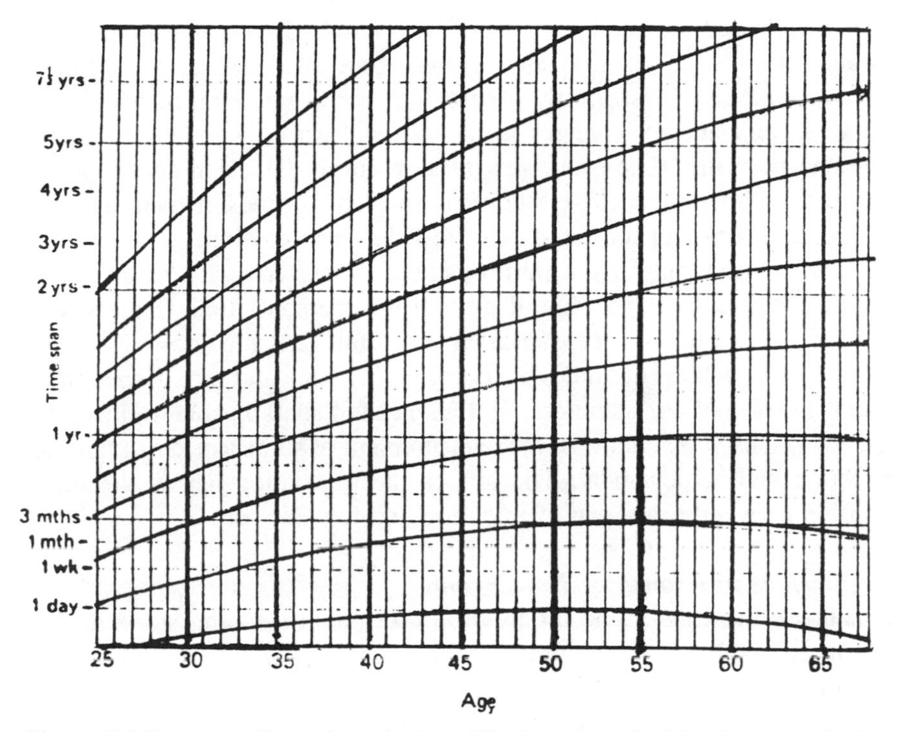

Diagram 9.3: Hypothesis of rate of growth of capability in adults, derived from learning profession curves.

fair and just at that time, his reply—X_2—will fall within ±3 percent of the figure derived from extrapolating the curve on which X_1 fell. A number of these X_1—X_2 extrapolations are shown in Diagram 9.3.[14] To get at the full significance of these regularities in people's expectations about reasonable progression in work and fair pay, it is necessary to recognize that the cumulative increases in what each individual would regard as a sound progression for himself in work level and related pay over these five-year periods range from 0 percent in the lower right-hand area—the older lower-pay group—to 60 percent or more in the upper left-hand area—the younger more rapidly rising group. These self-selected earning progressions are a financial expression of each individual's judgment of his own expected rate of growth in competence.

These judgments about progression in pay and competence have been checked against individuals' reactions to their actual progressions.

[14] These data were confirmed by Homa, who studied systematically the earning progression expectations of a population of 300 employees in six very different employment systems. Edna B. Homa, (1967), "The Inter-Relationship Among Payment and Capacity."

I have followed the careers of nearly two hundred men and women for periods of 14 to 22 years. Whenever one of them felt he was being fairly remunerated and satisfactorily progressed, the progression of his remuneration, adjusted for changes in the national earnings index, followed one and the same time-frame progression curve.[15] Deviations above or below this time-frame curve were experienced as times of overpayment or underpayment, of too rapid or of too slow progression.

Most important for our purposes here, however, is the picture of a strong consistency in the rate of growth in a person's time-frame when the foregoing findings are translated into time. Let me show how this translation is achieved, and, in so doing, bring time and time-frame back into the center of our attention. By this means I shall be able to illustrate the conditions which allow of the measurement of some aspects of human intention and achievement. The key to the translation lies in the relationship between time-span of intention and felt-fair pay.

Regularities in Growth of Time-Frame

I shall use the time-frame progression array as a stepping stone to the construction of my hypotheses about the patterns of growth of time-frame in individuals set out in Diagram 9.1.

The array of progression curves can be translated into an array of curves of progression in time-span by substituting the time-spans which correspond to the pay figures on the vertical axis of Diagram 9.3, but now reading these payment figures as felt-fair pay. The time-span scale can then itself be plotted along the vertical axis. The resulting curves from this transformation are shown in the accompanying Diagram 9.4.

The significance of these time-frame progression curves has been aided by following the careers of the group of nearly 200 people referred to above. The main finding has proved to be of the greatest importance for our analysis of the form of time. An individual who feels his competence to be fully employed—who feels that he is employed at full capacity—at a given time-span, at a given age will continue to feel fully employed, at full capacity, so long as the time-span at which he is employed progresses along that curve in Diagram 9.4 which contains the initial point of time-span and age.

Let me illustrate. A, at age 32 and at full capacity at 2-years time-span, felt at full capacity at age 38 at 3½-years time-span and at full capacity at age 45 at 5-years time-span. When at the age of 35 years he fell below this progression in time-span he felt underemployed. And when at the age of 40 years he was progressed above this curve in time-span

[15] Described in *Equitable Payment,* Chapter 10.

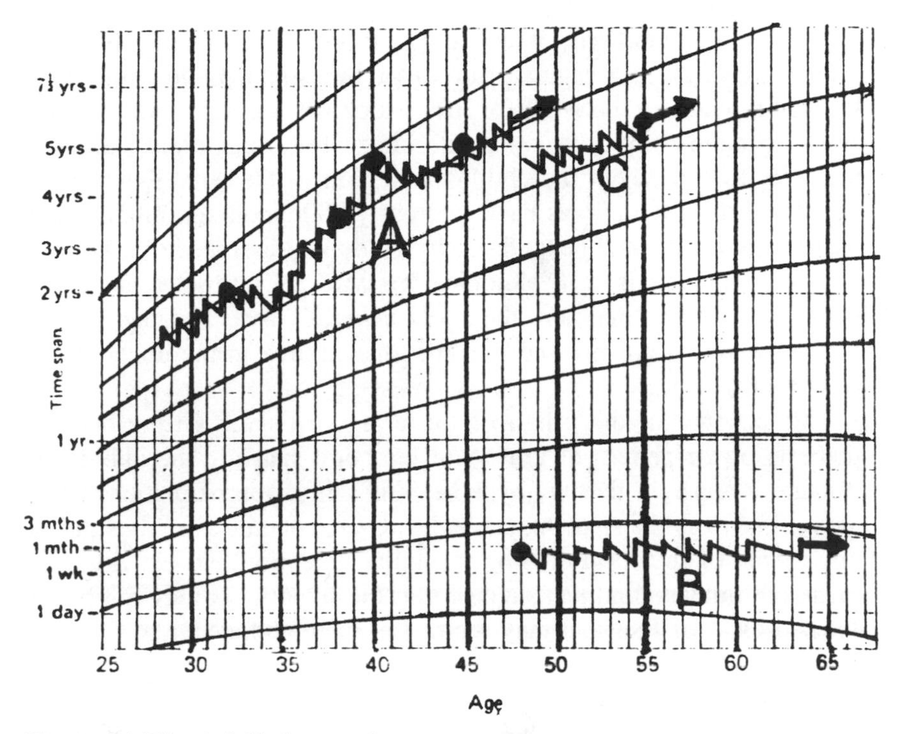

Diagram 9.4: Three individual progressions.

he felt the pinch and was unable to cope. B, at age 48 and at full capacity at one-month time-span, felt at full capacity at that same time-span through to retirement. He was not looking for any increase in weight of responsibility. By contrast C, at age 55 and at full capacity at 6-years time-span, was looking for continued growth in responsibility; he felt his competence to be steadily extending, contrary to the commonly held view that a person's competence is fully matured by the age of 45 to 48, and certainly contrary to the finding that IQ (for what that measure is worth so far as real work, real doing, as against acquisition of knowledge, are concerned) stops increasing by the age of 18 (*pace* IQ!).[16]

From these and similar findings over the past 25 years I have come to the conclusion that the array of time-span progression curves expresses a fundamental regularity in the pattern of growth of competence in adults, as measured by the time-frame—the maximum time-span of intention which the individual can muster in working activities, that is to say, in goal-directed activities.

[16] A full account of the findings and conclusions will be found in *Equitable Payment, Progression Handbook,* and *A General Theory of Bureaucracy.*

Given this hypothesis, it was but a short step to inquire into the possible existence of a regular pattern of time-span progression from childhood throughout the whole of life. I took the time-span progression curves, replotted them using a logarithmic scale for age along the horizontal axis to spread out the childhood region of the curves, and then extrapolated the adult curves back into early childhood and infancy and forward into old age and beyond death. To do so I assumed the sigmoid progression characteristic of biological growth, and drew the curves in Diagram 9.1. In that diagram, the adult array of curves in Diagram 9.4 will now be found compressed into the shaded area in the upper right-hand corner of the graph.

This account of the origin of the array of curves setting out the hypothesis of the pattern of growth of temporal space in the individual will make it apparent that the hypothesis is a first approximation. I felt it appropriate to publish it at this stage, however, not only as an illustration of my theme of the form of time in human intention and activity, but because preliminary findings suggest that the hypothesis may not be so farfetched as at first sight it may seem. Let me elaborate.

Time-Frame in Young Children

I have presented above some evidence for the regular growth of time-frame in adults. This evidence has been taken from adult engagement in employment work. Can there be any evidence for such a growth in young children? The size of time-frame would appear to be ascertainable in young children, for example, from the organized play or other activity which they can commit themselves to carry out—how long they can keep themselves occupied in doing a drawing, making a model, feeding, washing or dressing themselves, taking care of another child. As the child gets older, five- or ten-minute time-spans grow into half an hour, or hours, or days. Longer projects may be seen in school in doing homework, in private study, in carrying out a classroom project on one's own. Parents and teachers are well enough aware of the very great differences between children and pupils with respect to the maximum time-spans within which they can cope. The children with the longer time-spans at a given age tend to be identified as the more able and are judged to be likely to reach higher levels in their careers.

There is more systematic evidence embedded in experimental studies of children. This evidence is all the more telling because it comes from studies conducted for other purposes and with no conception of growth of time-frame in mind. Take, for example, some of the observational studies of infant feeding. There is evidence of initiation of feeding by the infant, in bursts of activity, with pauses, but between what are called gaps. These feeding episodes, combining bursts and pauses between gaps, are readily measurable in time. The episodes have been shown to increase in length of time with age.[17] Slow-motion films of these feeding episodes, with 1/10'' timer, are very dramatic.[18] They show the infant actively pursuing his feed, full of attention, alert, clinging, sucking, pausing, sucking. Then his demeanor suddenly changes: his facial muscles sag, eyes droop, and his expression becomes depressed; this phase is the gap. Then he suddenly perks up, and resumes active sucking for another episode. A striking feature of these data, however, lies in the fact that the analysis was carried out to discover whether there was on the average an increase in the length of episode with increase in age. This hypothesis was sustained. But it is possible from the raw data also to explore not only what happened on the average but also what happened to individual babies. Here my hypothesis would be that the longer the feeding episode of a baby at a given age, the more rapidly the length of feeding episode would increase with age. This hypothesis is sustained by the data, although the experimenters had not been concerned with individual differences.

These observations on the progression in length of time of infant-initiated and controlled feeding episodes are supported by other studies. Among these are the studies of infant recognition of mother as against a stranger.[19] Films of maintaining of eye-to-eye contact by the infant show a marked difference in the length of smiling and attention-getting episode when mother is presented as against when a strange woman is presented. The films were analyzed for how long over periods of 30 seconds the infant engaged in eye-to-eye contact. The results showed that two-month-old babies maintained an average contact of 18 seconds out of 30 with the mother, and of only 11 seconds with the stranger. In these infant studies we are, of course, dealing with episodes of seconds only. But Diagram 9.1 setting out our roughly constructed hypothesis about the growth of time-frame shows predicted maximum time-frames at this early stage in infancy of broadly the length of episode found in reality.

These data on growth of time-frame both in adults and in young children are of course very incomplete, and any conclusions must be tentative indeed. They are presented, however, in order to illustrate a

[17] See, for example, A. Fogel, (1977), ''Temporal Organisation in Mother-Infant Face-to-Face Interaction,'' pp. 119–152; R. A. Hinde and J. Herrmann, 1977), ''Frequencies, Durations, Derived Measures and their Correlations in Studying Dyadic and Triadic Relationships,'' pp. 19–46; and C. Trevarthen, (1977), ''Descriptive Analyses of Infant Communicative Behaviour,'' pp. 227–270.

[18] Trevarthen, op. cit.

[19] Trevarthen, op. cit.

theme; namely, that the problem of assessing that most elusive quality, a person's level of capability, may usefully be approached by conceptualizing it within a two-dimensional construction of time. Such an approach keeps a live linkage between level of capability and intentional or goal-directed behavior.

PART V

CONCLUSION

CHAPTER TEN

Psychological and Social Entities

The enigma of time has presented a series of riddles. First there was the question of the nature and existence of time and of whether it flowed from out of the future into the present to be left behind in the past. Such questions turn out to be verbal tricks once it is realized that time as a particular is an abstract ordering term, a term or state, or condition, or position, like form and place, and is not a concrete agent; hence it can be used to define the location of things and memories and intentions and the duration of processes, but it is not the kind of thing—like a river—which can flow or pass or go somewhere. It is a necessary conception for describing flowing or changing things, but is not itself changing or flowing. To speak of the flow of time itself is to reify it by making it into an agent. It would be like reifying space by conceiving it as holding its breath and standing stock-still at attention. This type of reification of time is a common enough human tendency, and, like all reification, leads to unnecessary confusion.

Then the question arose of whether or not time had direction, and if so, whether it was unidirectional—the arrow of time—or whether its direction was reversible. In considering this question, we came to the simple conclusion that the concept of time *per se* was not a directional concept any more than was space. What did have direction were the desires and intentions of human beings—the arrow of agency, in Stuart Hampshire's apt phrase. Desires and intentions are goal-directed and goals always have target completion times. They are thus both directional and "time-ful." They are directed toward getting or creating things which are capable of overcoming lacks, nothingness, absences, things missing, at some time or other, and usually as soon as may be. Hence the time in which they exist tends to take on their guise, and itself to appear directional.

Then there were the questions concerned with the many apparently different kinds of time: objective and subjective; real and imaginary;

proper and absolute; external and internal; physical and psychological; waking and dreaming; past, present, and future; flowing and still; clock and sensed; cyclical and serial; *chronos* and *kairos*. As we examined these different ideas, the differences became increasingly difficult to hold on to; they seemed to evaporate.

In the end we were left with the fact that there was only one concept of time as one concept of space—one univocal concept regardless of the setting, whether encompassing events in the mind of one person or events which are out there for all to see. This single concept refers to the way we organize our experiences of intentional transformation and successive achievement as against our experience of simultaneity. The sense of simultaneity gives rise to our concept of space—of things coexisting as points at a distance at the same time. The sense of memory and intention and achievement gives rise to our concept of time—of things continuing their identity while undergoing intentional transformation as points at a distance in succession. It is out of Will in the Kantian sense that our intuition of space and time arises in the phenomenal world.

We were still left, however, with, on the one hand, the problem of Heraclitus and Bergson seeing all as flux and durée, as Becoming rather than Being, while on the other hand there was the more secure world of things, of points at a distance, of bounded atomic structures, which had proved so useful as a starting point for the natural sciences, and from which vantage point the uncatchable, untrappable flux and durée had all the appurtenances of insubstantial mysticism.

I have approached this problem by taking what might seem to be the best of both worlds, and assuming a fundamental process of oscillation in human cognition between a field perception of the world and an atomic point-at-a-distance conception. At one moment the clear knowledge of atomic figure—whether things or ideas—is in conscious focus, against a preconscious and unconscious background field. Then the conscious atomic focus may fleetingly disappear, leaving us with a less organized and less focused feeling of awareness and sense of the surrounding world in which we are placed and which we contain within. It is this swinging back and forth between the two poles of the duality, each defining the other, and the consequent interpenetration and rubbing together of sense, awareness, and knowledge, which are the raw materials out of which meaning emerges, the materials whose integration *is* meaning.

In the course of these oscillations, time—or at least the processes, or changes, or continuities which we observe and which we relate to a concept of time—may appear in conscious focus. This focused atomic conception gives us our sense of succession, our sense that something was earlier and now is later—separated by a gap in succession which we call a length of time. In this perspective, time is verbalizable, measurable by clocks, something which we can "tell." In the background, our preconscious awareness and unconscious sense flows on, unpointed, unsharpened, unspoken, and unspeakable unless metamorphosed and born into the conscious world.

The other pole of the oscillation is the sense of flux or durée, or unformulated, unbounded, cloud-like, changing field—it is the continuous pole, the field that is always there. In the material world it is the field of electromagnetic phenomena, of gravity, in which centers of concentration may be described but in which the drawing of firm boundaries is a highly arbitrary act. And in the human world it is the field in which life goes on, in which discretion is exercised and decisions arrived at, the life of touch and feel and *nous* and judgment and clout and gumption, the life of the nonverbal and the pre-verbal, where feelings and justice and values and love and hate and desire and jealousy and disappointment and gratification and empathy and sympathy and affection and intention and the sense of self and other predominate, and intellect and logic give way to innate good sense (or bad sense) and the flow of unconscious rationality.

From this oscillating dual view of the world we derived a 2-dimensional analysis of the organization of univocal time—the temporal axis of succession (ts) and the temporal axis of intention (ti). This 2-dimensional construction solved for us the problem of the apparent contradiction between a seeming flow of future→present→past and a fixed earlier–later. Both are there in our analysis. We have, on the one hand, a dating axis, that of succession, which is not directional but which allows us to locate events in temporal relation to one another as temporal points at a distance. And on the other hand, we have a directional axis, the direction being given by humanly constructed goals, in which human intention builds upon its memory store and is pressed by it, in assessing its current lacks and desires, to strive to create that future which will satisfy desire. On this latter dimension we found our past–present–future existing as one present field of orientation and of action, in which the future is a present desire which shifts and changes under the impact of directed action by which the world is transformed. Out of this transformation something may come into existence which perhaps resembles the previously imagined goal—but it will come into existence out of constructive action and most certainly not as a preformed future somehow floating from out of a shrouding mist into the present.

The 2-dimensional analysis then was seen to be necessary only for the life world. A four $(3+1)$-dimensional world, the world of relativity theory, with the temporal axis of succession is quite sufficient for the world of physical things and processes. It is a cold material world (excluding, of course, the living world of the scientist himself as he engages in his research) devoid of intentions, devoid of feeling, devoid of direction, devoid of past, present, and future. The world of living things (including scientists), by contrast, is a warm-blooded world in which intention and intuition, desire, memory, and perception combine to create a continuously shifting past–present–future field. Here we require a five (3 + 2)-dimensional world view. The two worlds must not be confused with each other. They are both required if we are to cope with our justifiable and necessary concern with the physical matrix and the social matrix in which we are embedded.

Finally, we were led by our analysis into a consideration of the peculiar significance of the five (3 + 2)-dimensional world for locating human experience. What emerges is a picture of the psychological essence and size of the individual in terms of his time-frame and temporal domain. This time-frame took us squarely into the objective measurement of responsibility and of human competence in equal–ratio–length scale measurement by the same operation used in the natural sciences for the objective measurement of the properties of physical things. It also took us into a wider hypothesis about the growth of the individual's time-frame and temporal domain throughout life—a hypothesis with a possible predictive power which we have left as a preliminary proposition for further reflection and study.

Two-dimensional time seems an *a priori* given at the core of our organization of human experience, much as three-dimensional space seems an *a priori* given at the core of our organization of the physical world. It is some of these more general questions of the social and psychological world, human action, measurement, and the nature of an adequate science of human behavior, which I propose to speculate about in this final section of our exercise on the form of time, applying our 5-D analysis to see what difference it might make.

Entities and Processes in the 4 (3 + 1)-D Material World

With the development of the natural sciences from their first beginnings in Ancient Greece, our understanding of entities and processes has gone through at least three major phases which I shall call the static, the dynamic, and the relative. This language is only approximately accurate, but will serve my present purpose. In each of these phases, 3-D space is linked to time, but the linkage becomes tighter with each succeeding phase.

These changes make it appear as though the conceptions of space and time themselves undergo change, new conceptions emerging with each succeeding phase. The appearance, however, is deceptive. The same ideas of space and time—the same higher forms, the same categories, the same *a priori* givens, the same gestalten, the same constitutional imprint, the same cognitive organization—are applied. But the understanding of the nature of the material world changes, and so do the scientific description and formulation of that world. With these changes go changes in the conception of spatial and temporal relationships and how they are used in building increasingly complex models of the world with their growing predictive power.

The primitive starting conception is that of locating things in physical space. Time is handled separately in terms of dating seasons, in identifying how long it took or might take to get from here to there or from there to here, and in expressing views about eternal ancestry and the eternal future and their spirits and gods. The convention which is used to get some leverage or purchase on the problem is to abstract physical things spatially from the space–time plenum and to deal with them as if they are for the moment standing still and are unchanging, the dimension of time being used to date the static occurrence. This convention that static things exist—a convention which is too often taken as observed fact—is an integral part of the atomistic mode of conceptualizing the world.

By making this standard assumption about a conceptually static thing, we can proceed to describe it by giving its location and magnitude by reference to the x, y, z Cartesian coordinates, and can date its location by reference to the temporal axis of succession. By this exercise in abstraction we have identified our physical thing legitimately. I have used terms like locate, describe, identify, rather than existence, for these are the functions, along with measurement, for which we need and use the concepts of space and time.

Everyday life, as well as science from the Greeks to the Galilean revolution, deals with things and processes by getting a fix upon things, establishing their substantiality and extension by reference to their length, breadth, and depth. This quantification gives "the thing" which can then be weighed (if it is close enough and small enough), its specific gravity calculated, its movements recorded, and other features studied besides. There was no need for field theory for these purposes, nor for the purposes of geometry, trigonometry, and the associated survey and navigational procedures. Natural science was the science of statics.

The Galilean revolution was one in which the whole of man's relationship to time was transformed.[1] Galileo discovered how to bring time directly into conjunction with the spatial coordinates, and thereby was able to measure velocity and acceleration. As a result he could study directly the effects of gravitation and the characteristics of falling bodies, laying the foundation for Newtonian mechanics.

Central to Galileo's thinking was a profound change in outlook from the solely atomistic 4-D picture of the world with its seriatim cause–at–a–point followed in due sequence by an effect–at–a–point, to a picture in which falling objects, for example, could be understood in terms of their being part of a field of force to whose laws they were subject, and not carrying motion within themselves. It was this remarkable insight which introduced force and dynamics, not to replace statics but to open much wider the whole scientific perspective so as to embrace the study not only of physical processes but of heat-induced chemical processes, energy, pressure, work—in all of which time is an integral part of the equation. This dynamic space–time–force world is the one which Zuckerkandl argued is the more primitive cognitive organization.[2]

Then came the most recent phase, in which Einstein showed that not only can time be integrated with space to give dynamic fields of force, but the spatial and temporal characteristics of things and processes are themselves dependent upon space and time as jointly expressed in velocity. In this view, simultaneity, length, and duration take on different meanings in accord with the relative positions and velocities and direction of movement of observer and datum, and light has constant finite velocity.

In this current phase, space and time have become completely interthreaded, often described as a move from a 3-D to a 4-D conception of the world. That description of the change to a 4-D view is not true. The view of the physical world has always been conceived as 4-dimensional. That view was implicit in the theories of the Greeks, and made explicit by Descartes. What Einstein did was to build upon the fully developed 4-dimensional dynamic force–field conception initiated by Galileo and brought to fruition by Maxwell in electromagnetic field theory. In doing so, he forged space and time mathematically into a unified 4-D field in which the four variables were connected instead of separated as in Newtonian mechanics. It was then that Minkowski hit on the idea of constructing a 4-dimensional geometry.

What we have in the natural sciences, then, are three major orientations: the static-seriatim world; the dynamic field-of-force world; and

[1] This shift was described by Kurt Lewin in terms of the shift from Aristotelian to Galilean modes of thought. See Lewin, (1935), "The Conflict Between Aristotelian and Galilean Modes of Thought."

[2] In his distinction between space as place and space as force. Zuckerkandl, op. cit., Chapter 7.

the disturbing world of relativity. Each orientation uses the same three spatial dimensions and the same temporal dimension (axis of succession). Each orientation is useful in its own right depending upon whether the problem to be clarified occurs under static conditions, or is a dynamic force and process problem, or is concerned with velocities near the speed of light taken as a constant.

Finally, a seemingly paradoxical feature may be noted in this development, or perhaps a dialectical course of development, depending upon how you wish to look at it. Whitrow has remarked that the temporal characteristic of the world picture derived from relativity theory is that of a static world, a frozen world, a block time world. Relativity theory returns, in effect, to the more static pre-Galilean view of the world, but in a qualitatively different and much wider context—at a dialectically higher level.

The Deadening of Life by Imprisonment in a 4(3+1)-D Cell

One of the major dilemmas which faces anyone wishing to pursue the study of psychology and the social sciences is to maintain scientific rigor in method and precision in measurement without trivializing human behavior and losing the essential quality of humanity. For some time I had shared the common assumption that the cause of this dilemma was the general tendency to believe that scientific rigor required a material world outlook, and to adopt a spatialized cross-sectional perspective on human behavior rather than a time-filled cinematic perspective. This view, an adaptation of Bergson's reference to the natural philosopher's spatialization of time, is common enough among humanistic social scientists.

What seemed to be required, therefore, was to bring time in, to look at the growth and development of human beings and social institutions, to become dynamic; that would breathe life back into the human sciences, and take them out of the lifelessness of the spatialized natural science view of the world. I no longer believe this analysis of the problem to be correct. As I have related in the previous section, the natural sciences have never been spatialized; they have always taken cognizance of time—although the way in which time has been used has varied at different periods of scientific development.

It now seems to me that the key to the problem is to note that the only frame of reference that has been available thus far for scientific work is a 4-D frame of reference, in which the time dimension is the temporal axis of succession. This framework satisfies the conditions for understanding the material world. But to locate psychological and social phenomena in a 4-D world is to reduce the scientific view of human life to a mechanistic model of rats in a maze—three-dimensional physical organisms running in a three-dimensional maze in events called trials occurring at particular times the recording of which requires only the temporal axis of succession.[3] The idea that the rat has purpose, and that the experimenter is interpreting that purpose, can be suppressed because the poor rat cannot speak English, or whatever might be the native tongue of the experimenter. This type of physical 4-D description gives a sort of locational statement of where something happened or is happening. But it does not give any sense of the meaning to those involved of whatever might be happening, of their intent, their desires, their sense of duty or of responsibility, or perhaps of mere capriciousness.

The fact of the matter, of course, is that people can and do speak (and experimental psychologists and ethologists do assume that the activities of animals have identifiable meanings, an assumption strengthened by the recent experiments in teaching language to apes). Use is made of speech and meaning in conducting even the most highly controlled experiments with people or the most "hard data" types of survey in the social field.

It is too often assumed that to allow such a subjective and imprecise element as meaning into our studies must vitiate any possibility of the human sciences ever becoming exact sciences. According to this commonly held positivistic view, the moment that it is desired to introduce purpose, desire, memories, growth and development, will, reason, striving, logic, responsiveness, spontaneity, responsibility, humor and sadness, empathy, symbolism, and all the other accouterments of life, not just as detached 4-D physical behavioral data but as human phenomena with their full meaning to people, then at that moment all hope of exact science leaps out the window. Meaning and exact science are assumed to be mutually contradictory and incompatible.

Those social scientists and psychologists who would maintain that the positivist approach is the only respectably scientific and objective way to proceed, thus confuse the limited physical science 4-D convention with the whole of "reality" and impose it upon the whole of the world, physical and human alike. Nor, moreover, does it help to change within a 4-D framework to a seemingly more dynamic view by substituting, for example, a Bergsonian view of time as durée for the discontinuous time of the natural scientist. On its own, nothing scientific can be done with it. While seemingly allowing for more humanity, more purpose, it nevertheless, if used by itself, precludes all possibility of measurement, of quantification, and leads to a kind of vague unscientific view with almost

[3] Gregory Bateson has tellingly described this orientation as the "billiard ball" conception of behavior. Gregory Bateson, (1979), *Mind and Nature: A Necessary Unity.*

mystical undertones, a view which confirms the positivist in the soundness of his leaving meaning to one side in his search for a scientific understanding of human behavior.

I believe that this view that rigorous scientific method cannot be used to contribute to our understanding of human nature and social relationships with all their richness of meaning and intention, is correct so long as we constrain ourselves in a 4-D world construction. But it is incorrect the moment we shift to a 5-D world construction. In no way need scientific endeavor be unhuman. Scientific study is meant to aid our comprehension of our world. If in studying the human world science becomes unhuman, then it is no longer science. The significance of the 5-D context is that it gives the foundation for a rigorously quantitative scientific approach to human behavior and social process with the full richness of meaning and humanity and warmth kept in and not frozen out. I shall describe in the following sections why I think this seemingly paradoxical outcome occurs.

5(3+2)-D Definition of Human Entities

At the root of the problem to be resolved is the fundamental philosophical problem of definition: of how best to proceed to identify and to define what are variously referred to as entities, or individuals or objects or things; then to identify and to define objective properties of those entities; and then finally to identify and to define relative values or preferences attributed by individuals to those entities. This distinction between entities, objective properties of entities, and individual judgments about entities applies equally to physical entities and to psychological and social entities. It determines our modes of categorization or classification.[4] Finally, it is of the utmost importance for the theory and practice of measurement, especially for measurement and quantification in the human sciences. I shall, however, consider the definition of entities first, and then turn to objective properties and individually attributed judgments.

Let me illustrate the distinctions I shall make, by reference to physical entities and the physical sciences. In the case of material things and

[4] I do not intend a rigid or static view of categories. Categories are sets whose elements are defined in terms of the boundaries of the set. These boundaries, however, are alive in two senses. In the first place, they will always have anything from small to large differences in meaning to people, because each individual's experience with elements composing the set will have been different to a smaller or larger extent. And second, each time the term for an element in a set is used, the definition of the set has in practice been modified—perhaps in microscopic amount, but sometimes quite extensively; this point is nothing more than the obvious point that language is inevitably modified in use. I am taking this view for granted in my discussion and will not argue it critically. It applies equally to logical and mathematical sets, because logicians and mathematicians are also human beings who will have had widely varied and different experiences with the symbols used, and with the concepts of category or set as well.

processes, the first step in definition has been to abstract apparently fixed things at a point in time (ts), whether standing still or moving, and to locate them in three spatial dimensions. In my sense, of course, only one dimension of time—the ts axis—is used. It is a 3-D spatial cut against a 1-D (ts axis) temporal background. Armed with this physical visual discrimination or delineation, times of occurrence of physical events and processes could then be defined, and fields could be structured. The first cut into the spatio-temporal manifold—the first delineating abstraction—then, in defining physical things was a static cut: process and dynamics are then built up on this static foundation.

The basis of selection of entities for this process of abstraction is the familiar choice by use or function. We pick out what we need to pick out in order to live out our lives and relationships. We abstract all the things which matter to us. Our definitions are then definitions in terms of how we use the entities we abstract—so-called functional definition.

We quantify entities in their own right by enumeration, by counting them. Counting is the basis of accounting procedures. Counting comes easily in relation to the physical world, because different kinds of entities are so readily recognized through the concreteness of visual perception and of touch, and of 4-D organization of perception. We have the naive primitive belief that we can see those discrete objects just as they really are, that they really are out there just in the form in which we see them and can touch them. Few people ever become aware, or indeed need in practical terms to become aware, of the epistemological quicksand on which we live our daily lives. Be that as it may, in practice things are picked out in the context of any given culture, and defined in terms of use, function, pragmatics. We pick out, differentiate, recognize, and name those things which we choose to discriminate as things and name, choose to pick out, and treat as separate things, because for any of a thousand different reasons we find it useful to do so. Each culture has the things it wants to have and needs to have in order to function, and names for those things. The categorization of entities seems to develop readily, inextricably bound up with spontaneous learning of language in which all sorts of culturally determined categories are absorbed ready-made.

In the case of psychological and social entities, we also pick out those things which we use, which matter to us—like capability, committees, groups of various kinds, leaders, words, music, people, desires, ideas, castes, societies, tribes. But in picking out and identifying such entities, the same first cut which is used for abstracting material entities, the same initial abstraction, will not do. The difficulty of course is that we cannot get hold of them as fixed entities with length, breadth, and depth—we cannot perceive as things with a physical size, a thought, an attitude, or a tabu, or a value, or meaning of a word, or a need, or a social interaction, or a decision, or what makes one man a manager and another a salesman, or what makes ten people a team rather than a crowd, or a memory, or a marriage, or love, or anything having to do with people that matters emotionally to us.

By contrast with the initial static 4-D abstraction in defining the material world, the first step in locating, delineating, defining things in the human world must be the abstraction of social events or episodes, complete with their meaning for the participants, and with a beginning, a middle, and an ending. Such an abstraction, or first cut, is a 2-D temporal cut against a 3-D spatial backdrop. That is to say, the definition of an episode is an event which lasted from ts_m to ts_n, in which the intentions of the participants were directed to ti_1, ti_2, ti_3, . . . at various points in the event, the whole episode located at various spatial locations x_1, y_1, z_1, x_2, y_2 ,z_2, etc. The beginning of a human event can be established only in terms of an intention to do something; and the ending only in terms of a decision about completion or termination. In short, we require our 2-D formulation of time in order to come to grips with the most simple and primitive process of defining our psychological and social worlds.

This mode of defining things in the human world thus takes a 5-D perspective with a dynamic starting point which delineates phenomena which are saturated with the meanings and purposes of those concerned, in contrast to the physical world 4-D static starting point delineating things and processes which are in and of themselves devoid of any purpose or meaning. It is not that there is anything inherently better or worse in the one or the other starting point or first cut. It is that these are the ways in which the material world and the human world are phenomenally constructed and organized. The test of this proposition is a pragmatic one. If we take any other starting point, any other defining abstraction, then the natural sciences become magical, anthropomorphic, teleological, and the human sciences become mechanistic and drained of all meaning, or else they become mystical and extrasensory.

The objective elucidation of intentions is, of course, a difficult problem. But then so are the problems of understanding the functioning of the invisible world of quarks, black holes, and the rest of the esoteric armamentarium of physical thought. The point is that it is no use giving up because of difficulties inherent in the problem. It is precisely these problems which were discussed in Chapter 8. There it was suggested that time-frame could be assessed objectively by confining attention to time-spans of achievement related to goals which are socially established and agreed between two or more people. I now believe this point to be of central importance for the objective definition of intention.

Karl Popper has defined objective knowledge as knowledge which is socially shareable and shared. That definition can be directly applied to human intention. As I have shown in Chapter 8, if an intention has been agreed with another person, then that intention has *ipso facto* become objective. Moreover, the greater the socially agreed commitment to the intention, the greater is its objectivity. In this sense, the most objective types of social data are those which are legally defined whether by legislation, by judges, or by contract. That is to say, justiciable intentions *do* exist objectively, otherwise they could not be subjected to independent inquiry and judgment.

This limitation on our definition of intentionality turns out in fact to be no limitation at all—as far as scientific study is concerned, that is. For the essence of scientific understanding is that it is by its very nature, by its raison d'être, concerned with objectively shareable data, and objectively shareable data only. Objectivity is what science is about. It cannot deal with intentions so unconscious as to be unavailable to shared formulation. Nor, equally, can it deal with private intentions to which an individual is unwilling to commit himself objectively. In any case, most such noncommittal intentions can hardly be called intentions at all: they are mostly at best the merest shadows or imitations of intentions.

What then would such a 5-D definition look like? When we say a group of people is a football team, we assume not only that we shall find them involved from time to time in particular episodes called matches, with a beginning, a middle, and an end, but *also* that the members of the team would state themselves to be involved in those episodes with particular kinds of intention—to cooperate with one another to score goals, to keep opponents from scoring goals, and to achieve a general goal called winning. Without such episodes and intentions—no team.

Or when we say that two people are in love, we assume that we shall find them involved in a range of episodes or events which are part of being in love—items in the set—and that they have stated particular kinds of intention with respect to each other—events and intentions which will be very different from those for social and psychological things called hating, or envying, or voting, or working, or praying, all of which will have their own encompassing categories of episodes and intentions.

Committee meetings would be defined as a category of episodes involving certain numbers of people who have come together with the intention of resolving certain questions by previously agreed procedures, including majority voting; and a committee would be defined as a body of people who take part in such episodes. A vote in turn would be defined as a member of a category of episodes in which people debated and raised their hands or otherwise, with the intention of combining with others to win the day for some particular point of view. A manager-subordinate role relationship would be defined in terms of categories of interactions involving two persons in which one assigns goals to the other with the shared expressed expectation that one intended that a particular goal should be achieved by the other and the other stated publicly his or her intention to do so.

The definitions, in short, are functional definitions, except that categories of episode are the basic data, *and* the public intentions of the participants are always stated. Whatever might be the private intentions of the individuals in any given circumstance does not change the definition of the concepts. In a similar vein, I would define bureaucracy, parliaments, benefits, grading, attitudes, social classes, political parties, councils, power and authority, legitimation, competition, participation, different types of leadership, status, elected representative, employment, work, recreation, leisure, and so on, in terms of the particular categories of episode involved.

It is perhaps worth noting that this process whereby people objectify their intentions through public accord with others gives one means of explaining how we become aware of the mental life, or the mind, of others. G. H. Mead described how we become aware of the material world by the feeling of resistance of objects as we manipulate them with our hands to get them to go where we want.[5] In like fashion—and this Mead suggested as well—we get our most profound sense of the mind of others by the feeling of substantiality of resistance, of the will, intent, directionality of goal of the other as we strive to arrive at compacts of common intentionality, whether explicitly, or by emotional fumbling, pushing, and intuitive feel. It is in the interaction of intentions, the interweaving of goals, the sense of the other's intentions as obstacles or as facilitators, as frustrators or as satisfiers, or as merely irrelevant, as these are experienced in the course of goal-directed episodes, that we feel the psychological substantiality of the other person, with density, weight, a piece of reality to be shifted by psychological effort and influence, just as much as a door is in another sense a weight to be shifted. For people to synchronize in action is difficult: not to synchronize their watches—that is an easy use of the temporal axis of succession; but to synchronize the flow of their respective intentions, therein lies the toil from which we learn so much about one another.

Finally, it may be noted that there are more ways of categorizing

[5] G.H Mead, (1932), *The Philosophy of the Present*, pp. 124 ff; and (1934), *Mind, Self and Society*.

entities than by construction of simple unqualified entities. We not only have the category "book," but we can distinguish also between books with cloth bindings, paper bindings, and leather bindings. We can also distinguish between large, medium-sized, and small books; or between expensive, medium-priced, and bargain-basement books; or between useful, moderately useful, or useless books; and we can even distinguish between books that have been written, books that very probably will be written, and books that are very unlikely to be written.

The first of these categories might be termed categories of complex entities; that is to say, of combinations of simple entities, each of which exists in its own right, such as a book and its cover, or, to take another example, roles which are vacant and roles which are occupied by people. Take away the cover or the book, and you still have the book or the cover. Take away the role or the person, and you still have the person or the role. I shall not pursue this type of category further.

The second of these categories might be termed categories of entities qualified by their objective socially constructed properties; in the above example, books categorized according to size. If we wanted to refine our categories, as for example, putting the books away on different-sized shelves, we might categorize them as books up to 5 inches in height, those between 5 inches and 10 inches, and those over 10 inches. Categorization by objective properties raises a number of questions about the reality of properties, and about measurement, especially so in the case of human entities. It is such questions that I wish to consider in the next chapter.

The third and fourth of these categories—price and usefulness—both have to do with individual judgments or utilities attributed by people to entities. This attribution has to do with the attitudes which individual people have toward entities—with psychological attitudes which will vary from person to person and in the same person from time to time. This issue also I wish to examine in connection with procedures for getting the ratings of individuals of their own attitudes toward psychological and social entities.

The fifth of the categories—the likelihood that an entity exists or will come into existence—has to do with categorization in terms of ranges of probabilities. I shall not pursue this issue,[6] since I wish to focus upon the distinction between objective properties of entities and individual attitudes toward them, the theme to which I shall now turn.

[6] The question of probabilities and their assessment is in fact an important question in the problem of measurement which I shall consider. I feel, however, that the questions of measurement of properties and the valuation of attributes will take us as far as is appropriate in the present context.

CHAPTER ELEVEN

Quantification in the Human Sciences

Psychology and the social sciences have not as yet come close to achieving the remarkable directness and simplicity which attend measurement in the physical sciences. It will be argued in this chapter that a central reason for the difference between the two fields with respect to measurement is an epistemological one: it is concerned with the clarity of identification of entities and the objective properties of entities in physics; and the absence of such clear designations in the human sciences. How a shift to a 5-D framework might overcome some of these difficulties will be considered.

There are a number of characteristics of natural science measurement which contribute to its elegance and simplicity. The first of these is its ready use of extensive measurement, starting with the fundamental measurement of single properties of length, mass, and time; and its direct and easy construction of ratio scales from this extensive measurement. Then there is the fact that all the measures are reducible to a few primary dimensions, such as, for example, charge, temperature, mass, length, time, and angle, giving a coherent system of units. And finally, the combinations of primary dimensions always have the form of simple monomials, the various measures being relatable by physical laws.[1]

By contrast, in the human sciences extensive measurement is notoriously difficult and practically nonexistent. Most measurement is characterized by ordinal scaling or interval scaling; and when ratio scaling is established, it is mainly by means of complicated constructions either from polynomial conjoint measures or from differences of interval scales, rather than from simple extensive measurement.[2]

[1] These various conditions are comprehensively developed in D.H. Krantz, D.R. Luce, P. Suppes, and A. Tversky, (1971), *Foundations of Measurement: Vol. 1, Additive and Polynomial Representations*.

[2] Krantz et al., op. cit., and F. Roberts, (1979), *Measurement Theory*, Vol. 4 of *Encyclopedia of Mathematics and Its Applications*.

The simplicity of direct extensive measurement in physics derives from the primary abstraction and identification of physical entities as material objects extended in space; they are therefore locatable at any given point in time as objects with 3-spatial dimensions in a 3-dimensional Cartesian world, possessing the substantiality of mass. Entity and its fundamental measure dimensions exist in relation to each other and define each other. That simple picture has been modified in contemporary physics with the introduction of field theory from electricity and magnetism but, nevertheless, the fundamental referent of observation and of description remains the identifiable object or recorded datum.

By contrast, in psychology and the social sciences we have not yet made up our minds about the nature of our entities. We are by and large bogged down in 4-dimensional conceptions of atomistic psychic and social states or conditions, such as traits, or intelligence, or responses to stimuli, or social classes, or ill-defined bureaucracies.

Moreover, whereas the natural scientist is concerned to study only entities and constructed objective properties of those entities—objects and their length, density, elasticity, volume, charge, resistance, and some hundred other measurable properties—the human scientist is involved not only with psychological and social entities and their constructed objective properties but also with a third category of data, namely, the human preference for or valuation of all entities, and judgment of their utility. The separation of objectively constructed properties of psychic and social entities from individual preferences for entities, which differ from person to person, has not effectively been established.

A serious consequence of the situation in the human sciences is that it is not always clear whether price, or value, or judgment, or decision, or social class, or political attitude, or social benefit, or utility, or work are psychological or social entities to be counted, or objectively definable properties of entities to be measured, or preferential attitudes about entities to be rated. In the absence of clarity on this construction of knowledge, the problem of measurement becomes exceedingly complex, and unnecessarily so, as I shall endeavor to establish.

The main theme that I shall pursue is that many aspects of these problems of measurement can be resolved by shifting to a 5-dimensional framework for the human sciences. This shift will require the initial abstraction and identification of mental and social entities in terms of episodes with temporal duration. As shown in earlier chapters, the beginning and the planned ending of such episodes can be established and abstracted in terms of the objectifiable (social contracted) intentions of the persons involved, and the achieved end in terms of the termination of the episode by those persons. The significance of the primary identification of mental and social entities in terms of episodes with duration is that they are at one and the same time full of human purpose and feeling and not trivial, and readily subject to simple and direct extensive measurement of the physical science type, in terms of the objective properties of time of intended duration and time of actual duration of the episode.

It will be argued that the fundamental extensive measurement of time of duration of socially established goal-directed episodes may give the same kind of starting point for measurement of objective properties in the human sciences as the fundamental extensive measurement of length and mass give to the natural sciences.

The construction of such extensive measurement procedures then makes possible a clear distinction between the objective properties of human entities and individual preferences for those entities. The basis for this distinction lies in the fact that the objective measurement of properties demands the absolute zero of the positive real numbers of the ratio scale, whereas in the rating of preferences the zero is at most a point of indifference and negative real numbers are admissible.

In order to achieve the necessary precision of analysis, the term 'quantify' will be used to refer to the general process of attaching numbers to entities, to properties, and to preferences and utilities. The specific terms 'counting' entities, 'measuring' properties, and 'rating' preferences (including attitudes, utilities, and values), will then be used to refer to each separate mode of quantification.

What Is "Measurement"?

There have been continual attempts since the time of Durkheim and Fechner in the late nineteenth century to put psychology and the social sciences on the same footing as the natural sciences with respect to measurement: the comparison is usually with the fundamental physical process of measuring length or mass. There have been those like Campbell, the physicist and measurement theorist, who believed that it was not possible to achieve this aim because the prime condition of measurement could not be fulfilled, namely, that "a physical process of addition should be found [for the property in question] . . . . The differences between properties that are and those that are not capable of satisfactory addition is roughly that between quantities and qualities".[3]

Others, like Hays, simply note that little or no natural concatenation of psychological data has yet been achieved; he states, "As yet, the social

[3] N.R. Campbell, (1920), *Foundations of Science: The Philosophy of Theory and Experiment* (formerly entitled *Physics and the Elements*), p. 267.

and behavioral sciences lack clear agreement about those properties of man and his behavior that might be identified as fundamental *psychological* properties. Thus, for psychological measurement there exists no clear-cut parallel between such mathematical operations as addition and experimental psychological procedures. What exactly should one mean by psychological or behavioral addition? This problem may be solved some day; but at the present time only very primitive attempts have been made. It is safe to say that until the day comes and until psychologists and other behavioral scientists isolate and agree upon the fundamental measurement operations from which other measurement procedures will be derived and justified, the theory of measurement of psychological entities will be more incomplete and disorganized than measurement theory in the physical sciences."[4] Hays's argument is based on the fact that it is the "implicit psychological characteristics, abilities, wants, emotions, habits, attitudes, and perceptions, which are not directly observable but which must clearly lie behind their behaviors. In order to measure these psychological characteristics, one must infer their presence and their degree from behavior." He likens the psychologist's problem to that of the chemist trying to establish an—also implicit—atomic weight for an element.[5] As I shall try to show, I do not think the problem lies here.

Coombs and his colleagues also seem to agree that "No natural empirical addition arises in psychology."[6] Later in the same work they suggest that "the absence of a natural concatenation (or even bisection) operation in many areas of psychology, has led to the development of measurement models of a different kind."[7]

Krantz and his coauthors have given the most comprehensive statement of these difficulties. They have noted that "in the behavioral sciences, extensive measures are virtually non-existent, whereas interval scales do arise from various procedures, including conjoint measurement."[8] In their view "the attempt to apply extensive measures to the social sciences is beset with serious difficulties. In some instances, no operation is available; in others, the available operation either leads to trivial results or to a violation of the axioms. These difficulties have led to the development of other axiom systems as a basis for fundamental measurement in the social sciences, such as difference measurement and expected utility measurement. . . . Two psychological attributes, sub-

[4] W.L: Hays, (1967), *Quantification in Psychology*. California: Brooks/Cole, pp. 17–18.
[5] Ibid., p. 19.
[6] C. Coombs, R. Dawes and A. Tversky, (1970), *Mathematical Psychology*. p. 21.
[7] Ibid., p. 25.
[8] Krantz et al., op. cit., p. 517.

jective probability and risk, are exceptional in that they appear to be extensively measurable."[9]

It is to this question of the seeming unavailability of simple and direct extensive measurement in the human sciences, except perhaps under very exceptional circumstances, that this chapter addresses itself. The mixture of views described is characteristic of the general outlook today. There is an inclination to agree with S. S. Stevens that physics-type extensive ratio-scale measurement is achievable in psychophysics but not in the broad area of psychological behavior as a whole.

The difficulties and uncertainties have three main sources: first, a failure to define what is being measured, so that no clear comparisons are possible; second, a failure to note that whereas physics deals with entities and the objectively definable properties of entities, psychology and the social sciences deal with both these and with individual judgments about entities as well (including values, attitudes, and preferences), getting tangled up between these objectively definable properties and individual judgments; and third, a bogging-down in the world model of physics which excludes purpose and intention, whereas a world model which includes purpose and intention is called for where human activity is concerned, and when used, makes possible the objective measurement of properties. I shall deal with each point in turn.

The unclarity over definition, and the profound practical significance of that unclarity, can be shown in the various definitions of measurement which are used, for example, by the authors I have mentioned above. They are typical of the general outlook. S. S. Stevens, for example, defines measurement as the "assignment of numbers to objects according to rules."[10] This sounds all right, until one notes that Coombs and his colleagues refer measurement not to objects but to properties. As they put it, "the process by which the scientist represents properties by numbers is called *measurement*."[11] But later on they bring in objects and scaling as well as properties and measurement: "The actual process of assigning numbers to objects, or properties, is called scaling."[12]

This confusion between measurement as concerned with objects or properties or both has serious consequences, as I hope to show in a moment. It is illustrated in Hays, who defines measurement at first in such terms as "certain properties of the things studied by the scientist

[9] Ibid., p. 124.

[10] S. S. Stevens, (1951), "Mathematics, Measurement and Psychophysics," in Stevens (ed.) *Handbook of Experimental Psychology*.

[11] Coombs, Dawes and Tversky, op. cit., p. 7.

[12] Ibid., p. 31.

are measured, or given numerical values.''[13] But then he uses the term 'measurement' to refer simply to classification, as for example, to such processes as ''seeing whether a certain solution is an acid or a base''[14] by inserting a piece of litmus paper into it in order to classify it. Thus he writes that ''the assignment of objects of observation to categories according to some classifying scheme and following some specified rules of procedure is measurement at its simplest and most primitive level.''[15] By the same token, he argues that a psychiatrist diagnostician determining whether or not to classify a patient as psychotic is measuring the patient. Then he applies the term 'measurement' to the placing of runners in a race in the order, 1, 2, and 3,[16] meaning, I suppose, that we measured the first, the second, and the third runner by means of ranking order.

This indiscriminate use of the term 'measurement' to refer to such disparate procedures as assigning numbers to objects, or to objective properties, or to individual preferences, and to assign objects or people to categories, is characteristic of much writing in psychology. Yet in everyday practice the term 'measurement' is not used in this way. Certainly the physicist does not use it so. He distinguishes between two quantitative processes: one is the process of enumeration, of counting how many entities there are, counting their frequency of occurrence perhaps, or counting how many stars of a certain type there are; and a second is the process of measuring the properties of his entities, their objectively definable properties—their length, weight, temperature, wavelength of emitted light, hardness, elasticity, horsepower, electrical resistance, and so on. He does not refer to them both as 'measurement.' In physics, measurement theory is concerned solely with the assignment of numbers to the properties of entities (objects or events), and not to the entities themselves, or to classification, or to utility, or value or other attributes.

Krantz et al. have substantially clarified the definition of measurement. They define it as the existence of a well-defined homomorphism between an empirical and a numerical relational structure. The properties of the empirical structure must be shown to justify a numerical representation, and that proof constitutes a representational theorem.

Extensive measurement yielding ratio scales is based on the homomorphism of the empirical structure of a set of elements with an ordering relation of some kind and additive concatenation, and the numerical structure of the positive real numbers, the relation of greater than, and

[13] Hays, op. cit., p. 1.
[14] Ibid., p. 5.
[15] Ibid., p. 7
[16] Ibid., p. 9.

numerical addition; that is to say, the homomorphism of $\langle A, \succ, \circ \rangle$ into $\langle Re^+, >, + \rangle$.[17]

Some ratio scales can also be derived by means other than extensive measurement—for example, by polynomial conjoint measurement and by taking differences of equal interval scales.[18] But these methods of measurement depend on taking more than one attribute at a time, thus excluding the great simplicity of the extensive measurement of single properties in physics.

But even Krantz refers to all homomorphisms of the above type as measurement, including those whose data are based upon individual ratings of preferences as against data based upon the objective construction of properties. Despite the parsimony and clarity of their definition of measurement to cover all attributes, both properties and preferences, I shall argue that the distinction between these two types of attribute should be sustained, and shall show why.

Quantification: Counting of Entities, Measurement of Properties, Rating of Preferences—Some Definitions

I propose that we should use the following terminology so that we can discriminate between significantly different processes of quantification, each being of importance in its own right. It is consistent with everyday usage.

The usage to which I am referring is the everyday sense of the difference between counting, measurement, rating or evaluation, and probabilistic judgment. When we ask someone to assign numbers to a pile of carpets in the sense of how many there are, we do not ask him to measure the number of carpets but to count them. If, however, we want to know whether the carpets will fit into a room, we do ask the person to measure the carpets, it usually being tacitly understood from the context that we mean him to measure not "the carpets" but the length and width of the carpets, by means of an objective yardstick. If, however, we want to know whether that person thinks the carpets are worth their asking price, we do not ask him to measure their value, we ask him to value them or to rate their value or evaluate their relative worth by giving us his individual or personal judgment of how he would rate them in relation to other similar carpets at different prices. If, finally, we want to know whether the carpets will wear well in the conditions under which they will be used, we do not ask to have their lasting qualities measured, we ask for a judgment of the probability of length of life (a judgment

[17] Krantz et al., op. cit., pp. 9 to 12 and pp. 71ff.
[18] Ibid., p. 518

which might conceivably be aided by measuring certain objectively definable properties, such as the toughness of the jute backing, and taking such measures into account in making the probabilistic judgment).

Even in this simple example, the distinction between counted entity and objective measurement of property would in fact have to be made explicit if the carpet were to be ordered. The distinction in such a case would ordinarily be made in terms of 'quantity' and 'size'; that is to say, for example, counting out an order for one hundred carpets (counted entities, commonly called 'quantity') measuring 15 feet by 10 feet (measured properties, commonly called 'size' in such cases).

Now, the question which confronts us is whether or not to use the term 'measurement' for all four procedures, as current measurement theory says ought to be done. It would mean a usage in which we would be forced to say, if we were to be consistent, "measure the number of carpets, measure their length and width, measure their value in relation to price, and measure the probability of their lasting." I cannot find that such a usage is employed by anyone, even by measurement theorists.

I propose, therefore, to use the following terms. I am using the term *entity* to comprehend objects, fields, events, episodes, including statements, which are objective in the sense of socially shareable.

I shall shorten the phrase 'objectively definable property' to *property* to refer to objectively specifiable features of an entity. These features, such as length or hardness of things, or level of work, or length (temporal) of a contract, are objective in the sense that the way they can be observed, and eventually measured in magnitude, can be stated using a socially established instrument which permits of social agreement regarding the data in the sense that everyone using the instrument will arrive at the same measured quantity. Such properties give the appearance of being independent of people and consequently inherent in the entity; but in fact they arise from people's interaction with the entity, observation of objectifiable data, and organization of the data in terms of constructs called properties.

The phrase 'individual judgments and preferences about entities' (including utilities, attitudes, and values) will be shortened to *preferences* to refer to the different opinions which different individuals might have about an entity such as beauty, or benefit, or usefulness, or other valuation as compared with other entities, in terms indicating which is preferred. Preferences are thus individual and personal feelings about an entity. They can never be the subject of objectifiable social norms shared by everyone, but only of norms which are individual. Their outstanding characteristic is that they vary from person to person, and in the same person from time to time, in accord with personal feeling. I shall continue to keep the concept of *probability* separate from preferences, including utility, although I shall not deal with probability at any length.[19]

I shall use *quantification* to refer to the general process of assigning numbers to entities, to properties, to preferences, and to probabilities. I shall then adopt the usage: *counting* of the number of entities, *measurement* of the ojectively definable properties of entities, *rating* by individuals of the preferences (or utility) they attribute to entities,[20] and *quantified judgment* of the probability of outcome of the as yet unknown.

This terminological specification is critical for the human sciences, and for the demonstration of the importance of the 2-D analysis of time in establishing the nature of entities, properties, judgments, and probabilities in these sciences. In everyday usage, and in the natural sciences, the context will ordinarily let us know what is meant so that language can wander all over from measure, to count, to state the quantity or the number required, or the size, or the weight, or the price, or the size of the bargain. In the psychological and social sciences, however, the context is far from well enough established to allow of such linguistic looseness.

Some Consequences of Confusing Properties with Preferences

Physical scientists do not know it, nor do they have to know it, but when they qualify their entities they do so in terms of measuring their properties but do not get involved in rating preferences. The story in the natural sciences has of course not always been so straightforward. The natural scientist also, at one stage, was beset by confusion between objective properties of entities and anthropomorphic and magical attributions to entities. Indeed, one of the problems of objective definition of properties is precisely that of how to avoid such anthropomorphic attributions to physical things and to be able to know objectively that one is doing so. This process lies at the heart of the separation of science from magic. For the magical lies in the individual attribution of values to physical

[19] The problems of the nature of probability judgments and their quantification are sensibly discussed by Phillips. He draws attention to the inadequacies of both the psychophysical models and the information-processing models for explaining probability judgments. He puts forward an alternative model based upon the conception of judgment processes carried on within a hierarchy of mental structures. This view is congruent with the discontinuity theory of levels of abstraction that I have outlined elsewhere (*A General Theory of Bureaucracy*. London and Exeter, New Hampshire: Heinemann Educational Books Ltd., (1976)). It may be noted that probability statements may be applied to the likely truth of retrodictive, dictive, and predictive statements about entities, properties, or rating. L. D. Phillips, (1980), "Generation Theory," *Working Paper 80.1*, Brunel University Decision Analysis Unit.

[20] This usage of rating allows the term *evaluation* to cover the total process of assessing the outcome and usefulness of given programs. That is to say, a full evaluation might include: a counting of certain effects; measurement of certain properties of the program and its effects; ratings of the social value of the program; and probabilistic judgments of the effects of permanent implementation of a pilot-tested program.

things as though these human values were properties of the things themselves. Part of the art of the magician is to make us believe that physical objects have taken on life. This primitive mixing of objective properties and attributed values, of the 'real' and the 'unreal,' of science and magic, persists widely in the everyday world of belief in ghosts, in miasmatic theories of illness, in magical cures, in miracles, in misperceiving things seen in the sky as UFOs, in fears of the uncanny.

In natural science itself it has not always been easy to find the way through to the identification of objective properties. It took Galileo, for instance, to refute the Aristotelean conception that motion could occur only by the continued application of external forces and could not be inherent in things, and so to formulate his law of inertia.[21] Contrariwise, one of the major steps forward in chemistry was the recognition that heat is a property of molecular activity and not a calorific substance, namely phlogiston, passing in its own right from one object to another. The overcoming of the problem of constructing properties lay in the great explosion of discoveries of scientific instruments which could be used for the objective measurement of such properties—the calorimeters, spectroscopes, thermometers, ohmeters, barometers, telescopes, microscopes, wattmeters, during the eighteenth and nineteenth centuries, and the enormously sophisticated instruments connected with research into the atomic world in the twentieth century.

In contrast to the modern natural scientist, human scientists have by and large tended to focus their activities upon the study of attitudes and judgments, and the processes of attributing individual values, preferences, and utilities to things. By comparison there is some but not much preoccupation with objectively constructed properties of psychological and social entities, which might constitute empirical relational structures which can be represented by the numerical relational structure $\langle Re^+, >, +\rangle$. Many serious consequences flow from this situation. There is the inhibition of the discovery of objectively definable properties through neglect of the necessary searching, or through searching in the wrong way and in the wrong place. Then there are many unrealistic arguments about values, arguments which are unrealistic and unresolvable in that they are really about objectively measurable properties and could be settled readily and effectively if such measurement were used. And then there is the turning away from purposefully directed mental activity itself, characteristic of behaviorism and other limited positivistic approaches to scientific method.

There are many studies and propositions, for example, in which

[21] The law of continuous motion of objects unless decelerated or accelerated by external forces.

preferences and properties are confused in the sense of each being mistaken for the other. The consequences can be socially and politically disruptive. As shown in Chapter 7, for example, there has been the failure to recognize that the level of responsibility in employment roles can be construed as an objective property of the work in the role, extensively measurable on a ratio scale by time-span of discretion, and is not merely a value to be attributed by job evaluation rating procedures[22] or by judgments obtained by sociological surveys of the relative status ascribed to various occupational titles.[23]

Then there is the equally common misconception that an employee's performance—which can only in fact be an attribute rated by a manager—is a property of the employee's output to be measured objectively in terms of counting that output. Such thinking has been an effective obstacle in the way of the development of sound systems of differential reward in democratic industrial societies.

Then there are the so-called objective cost-benefit analyses which have become so widespread. In the first place the notion that cost is an objective property of things is a misplaced idea. Cost is in fact a subtle and complex and elusive mixture of material purchases, labor input, and investment—in the final analysis, the totality of labor value and investment value—all of which are attributed values based upon preference ratings.[24] And in the second place there is the equally unrealistic—I think it would not be too strong in this case to call it absurd—idea that social benefit can be objectively measured, "scientifically evaluated" is the phrase, that it is an objectifiable property of the outcome of a program rather than an attributed value to be judged by policy-makers and those in receipt of the benefit of the program.[25] Social scientists can be of

[22] So-called job or work measurement or job evaluation sometimes means a counting of outputs; sometimes a counting of the various movements of the employee and a timing of those movements; sometimes a rating of performance value; sometimes a rating of responsibility; sometimes a rating of the social status of the job title. And the relative advantages of these various processes are argued about as though they were alternative methods of doing the same thing, namely "measuring the job," for the same purpose, namely paying for it. But, unfortunately, it rarely means measurement of the level of work in the role.

[23] The results of such surveys are one of the major attributed values built into our ill-conceived sociological conceptions of social class.

[24] The mythology of costing and pricing is discussed in Wilfred Brown and Elliott Jaques, (1964), *Product Analysis Pricing*. London and Exeter, New Hampshire: Heinemann Educational Books Ltd.

[25] For what would it mean to measure the benefit (a common enough phrase, in truth)? It could mean to count the particular outputs from a pilot social program: for example, in an alcohol detoxification program, a counting and statistical analysis of so many alcoholics treated on so many occasions, with an average frequency of so many treatments per person. Or it could mean to measure (or to try to measure) objectively certain inherent properties of the program or its outcomes, such as the low intensity of withdrawal symptoms in individuals. Or it could mean to rate the social value of the program to the alcoholics, and their families, and the community. Or it could mean to judge in quantified terms the probability that implementation of the program permanently on a wider scale might help to deal with the problem of alcoholism in the community. Or it could mean all four. It is important to know which.

assistance in gathering such attributed value judgments and, as many decision analysts do, in helping the policy-makers to arrive at their own best judgments. But objectively measured benefits as though they could be constructed as objectively definable properties and extensively measured, never!

It is not difficult to list other examples of such confusion between individually attributed values and objectively shareable properties—assessments of the outcome of psychological treatment, productivity bargaining, studies of quality of life, modes of changing attitudes, leadership training, historical materialism, the characteristics of groups—the list is readily enough lengthened. The question is what to do about it. My reply to that question is to turn to a model for the human world which includes purpose and intention. In order to demonstrate how and why such a step can transform the problems of measurement and rating in the human sciences, by pointing to a wide range of readily accessible empirical relational structures, amenable to extensive measurement, let me turn to the fundamental question of the measurement of properties of entities of any kind.

Absolute Zero-Based Concatenation in Measurement

It has long been noted that there are a very few—perhaps three—so-called fundamental methods of measurement in physics. These fundamental measures are length, weight, and time. The sense in which they are most commonly seen as fundamental, rather than derived, is that each is measured in terms of itself.[26] Length is measured by a concatenation of entities of equal length (say sticks); weight is measured by a concatenation of entities of equal weights; time is measured by a concatenation of equal time intervals, such as swings of a pendulum.

In addition to the concept of fundamental measurement, it may prove useful to recognize that there is an important sense in which length measurement is primary. For both weight and time scales can be reduced to readings upon a length scale (and this is commonly done in our measuring instruments) by any of several means in each case; for example, by measuring time in terms of the rotation of a clock-hand about a fixed center point, or measuring weight by equilibrium in rotation about a fixed fulcrum point, both of which can be reduced to readings in length measurement.

The significance of being able to reduce all measures to readings on a length scale is that the length scale gives a direct physical representation

[26] N. R. Campbell, op. cit., uses such a definition (p. 378), although he uses this categorization as a convenience rather than as the expression of some principle; and Krantz et al., op. cit., follow this usage (p. 456).

of the scale of positive real numbers—including both the continuity of the stick and the discontinuity of the numbered marks on the stick. I shall use this fact in demonstrating that equally fundamental measurement, using 2-D time, is not only possible in the human sciences, but that such measures abound naturally—they are all around us, staring us in the face, if we learn to see them. Let us, therefore, scrutinize this process of additive concatenation more closely, using fundamental length measurement for the purpose.

The process of extensive measurement by additive concatenation, as it is commonly described, is to take any unit length and then lay out a series of such lengths, as illustrated in Figure 1. This physical procedure

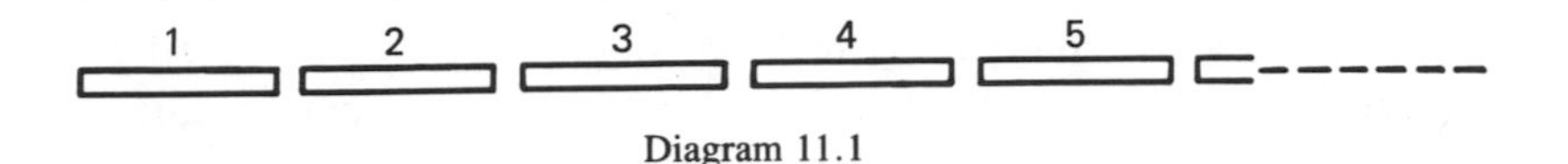

Diagram 11.1

establishes an empirical relational structure of the form $\langle A, >, \circ \rangle$ which is a homomorphism into the numerical relational structure $\langle Re^{+}, >, +\rangle$. The empirical elements can be mapped onto the scale of positive real numbers. In so doing, the same operations of addition can be carried out by arithmetic on the numerical scale, as by the more cumbersome laying out and counting of more and more sticks along the ground (i.e., 4 sticks added to 3 sticks will count up to 7 sticks, the same result as the arithmetical operation $3+4$).

There is a feature of this process which must be noted. It will be apparent that the zero in the numerical relational structure means real zero. But the equivalent zero on the sticks has to be explicated as well.

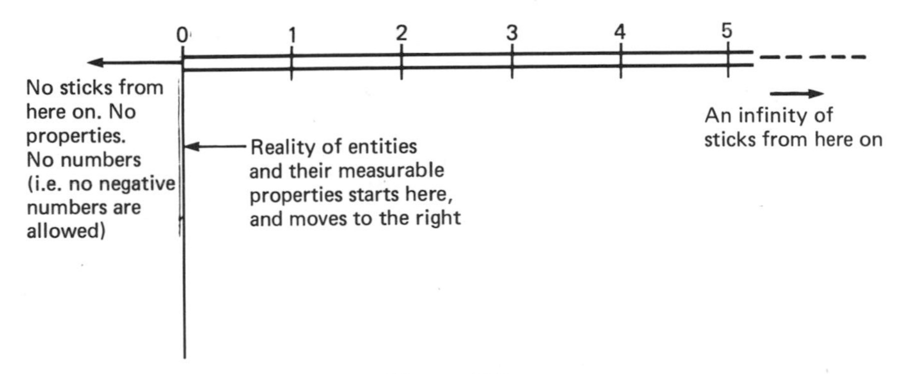

Diagram 11.2

It is at the left-hand end of the left-hand stick. It means that the sticks are asymmetrical in position, reading from left to right. And the significance of the zero point at the left end of the left stick is that there is absolutely no more property of length to the left of that point, and absolutely no more stick. It is a real zero: it means no more sticks; zero sticks; nil sticks; nothing; no more length, and certainly no negative length. It is an absolute zero: everything stops here.

It is, of course, the existence of this absolute zero, both in the empirical physical concatenation process and in the numerical positive real number model, which squeezes out of this process an equal-ratio scale, one in which $12=4\times3$ as well as $9+3$; that is to say, a scale on which you can multiply and divide as well as add and subtract.

Finally, it is to be noted that the length scale—the physical scale, that is—has a most interesting feature, one which is possessed by no other measuring instrument. This feature is that the physical concatenation process can be represented on one stick, a length rule. It is an actual physical expression, a physical model, of the numerical positive real number scale,[27] illustrated in the accompanying drawing of a yardstick divided into feet and inches. It could be further divided into 1/10″, or 1/100″, or even smaller segments. But no matter how finely it is divided, it retains two immediately and directly perceivable features: it is both finitely continuous and infinitely discontinuous. It is a physical manifestation of both the continuous thread and the discontinuous beads along the thread, used to describe the number scale. In short, this type of rule is a complete empirical relational structure which gives an objective physical expression of the numerical structure of positive real numbers.

It is this feature of containing both continuity and discontinuity which allows the yardstick to be used to measure a 1′3″-long stick as a continuous

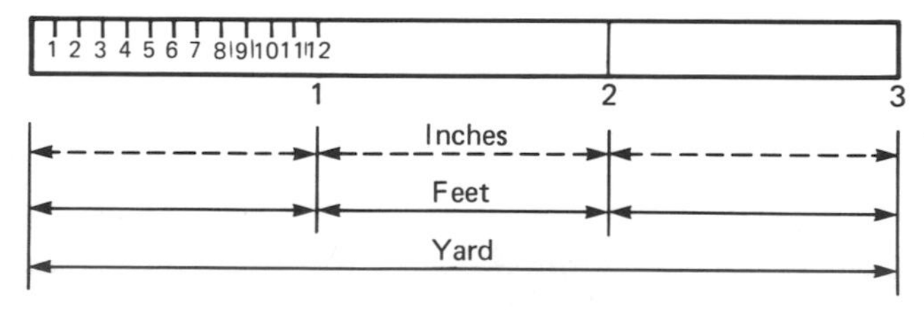

Diagram 11.3

[27] I have always felt that the first person to mark off an absolute-zero-based equal-ratio scale and use it for measuring ought to have a place in the Hall of Fame between those other two well-known stalwarts of the development of civilization, the person who first used fire and the inventor of the wheel.

1′3″ enclosed within the discontinuous end-points at 0 at one end and at 1′3″ at the other—and to perceive this total state of affairs all at one and the same time. Extensive yardstick measurement of this kind is absolute-zero-based length measurement.

Let us now turn to measurement and rating in the human sciences. I propose to show how we can use extensive measurement to produce ratio scales of the properties of human entities. I shall also show that as far as rating of preferences is concerned, it is possible to generate ordinal and equal-interval scales but not equal-ratio scales. I shall advance a possible explanation for this fundamental limitation with respect to the rating of preferences.

Measurement of Properties of Entities in the Human World

If against this analysis of measurement in the natural sciences we move to measurement in the human sciences, a whole world of natural extensive measurement opens up—measurement that is as straightforward and precise as the fundamental measurement of length and mass of physical objects, and most often with no problem of observer error.[28] To see that world, however, we shall have to use a world model which retains purpose and intent. Having done so, we can then measure directly in terms of time the intended or planned or agreed times to be allowed for gaining particular objectives as shown in the descriptions of objectively established goal-directed episodes to be found in Chapter 7; and we can equally measure the time taken for the achievement of those objectives. I refer to such obvious entities and properties as: the negotiated length of a contract; the time by which someone publicly commits himself to doing something; the predicted and actual duration of an economic boom or recession, or the time-scale of economic cycles; the potential 5-year life of a parliament; the planned duration of an engagement; the duration of a football game; the periods chosen for planning cycles.

These lengths—lengths of time, that is—are true measures of significant properties. Take, for example, the length of a contract; if a builder commits himself to finishing a particular job in 3 months rather than 6 months, that is a significant and objectively measurable difference. Equally, whether a football player, or a consultant, or a research unit, obtains a 1-year, or a 3-year, or a 4-year, or a 10-year contract is a matter of great substance. And note, to obtain a 4-year contract is to have a

[28] I shall not go into the question of errors of measurement—either observer error or errors due to shortcomings in instruments such as, for example, wear on the knife edge of a balance. But it may be of interest to note that extensive measurement of a socially contracted agreement to complete an assignment in, say, two days does not allow of any observer or instrumental error. A publicly agreed 2 days is 2 days.

contract that is twice the length of a 2-year contract. Or to add 2 years to a 3-year contract gives a 5-year contract. We are dealing here with simple and direct extensive measurement and ratio scaling. A contract of zero time duration is a zero contract: it does not exist.

The same features hold true of all other measurement of the property of the episodes which are the manifestations of psychological and social entities. It is no accident that time, duration, should enter in this way. The temporal length of an episode states its primary existential quality in the same sense that spatial length does for a physical object. The intended and the actual duration of an episode—whether a stated intention or a retrospectively constructed succession—yield two of its basic properties. Unlimited numbers of such natural measurements are available simply for the observing.[29]

It may at first sight seem trivial and inconsequential to tie the fundamental measurement of properties of human entities simply to their intended and achieved durations. I do not believe that it is trivial. Such measurement is of no less consequence than is the fact that spatial length is the fundamental mode of measurement of physical entities. There is no special significance in the 3-D spatial length of something: it is what follows from such measurement that counts.

The same holds true for temporal length measurement of human entities. It gives a foundation measure which is not only akin to natural science measurement but is, I believe, the direct homology. As such it

[29] N.R.Campbell demonstrates the difficulty of dealing fully with time solely from the physicist's $4(3+1)-D$ perspective. He continued to bracket length, weight, and time as the three fundamental measures, even though he remained unsure of time. His intuition told him one thing, but his limited $4-D$ conception made him doubt his judgment.

In his discussion of the conception of time, he writes: ''All temporal conceptions depend on the immediate judgments of 'before,' 'after,' and 'simultaneous with','' thus committing himself exclusively to the temporal axis of succession, as a physicist should and must do.

But then he adds: ''all physical judgments can be ordered in respect of time. But the establishment of this order does not lead immediately to a system of measuring 'time,' because no satisfactory method of addition can be found. There is no rule whereby I can combine an event that occurs at 3 a.m. with one that occurs at 6 a.m. to make an event which occurs at 9 a.m. (...it is impossible to give any definition of the term beginning an 'event is...').'' N.R.Campbell, op.cit.,p. 550.

Campbell's difficulty arises from the fact that there is no such thing as an event which occurs *at* 6 a.m. or *at* 9 a.m. Events can occur only *between* 5 a.m. and 6 a.m. or *between* 8:59 a.m. and 9 a.m. That is to say, any event which exists in the sense that its properties are extensively measurable, must have duration. And durations can be concatenated. An event which takes one hour can be added end to end to another event which takes 2 hours, to give a total of 3 hours of events; just as in physics you can place two sticks end to end and total their measured length.

I believe that Campbell failed to see this simple point because he had intuitively mixed up his physical event time of succession with time of intention and human purpose. From this latter point of view, his example could read, ''Starting now at 3 a.m., instead of trying to finish at 6 a.m. take another 3 hours and try to finish by 9 a.m.,'' and that statement certainly does contain the temporal addition necessitated by the increase in time–span of intention for achieving the goal. But the addition is now on the temporal axis of intention.

is scientifically impeccable. But, because it is a goal-directed temporal measure, this rigorously scientific measure cuts right into the heart of human life: into promises to do something by a certain time, and the keeping or not of those promises; into wishes or longings that something might happen by a certain time, and the realization or not of those wishes or longings; into the anxieties that accompany the commitment to produce something for someone, and the awareness that time is running out and the commitment may not be fulfilled; into the sense of order which derives from things running as planned, in accord with schedule; into the disappointment, or anger or rage, when someone else fails to deliver on time; into the judgments about people's being reliable or unreliable. It is things like these that are at stake in the marriage of intention and achievement in human life: and it is precisely this interplay of intention and achievement which is objectively measurable in terms of time.

Derived Measurement of Properties

When we come to derived measurements of properties, however, the situation, as in the natural sciences, becomes more complex. It calls for the development and construction of special instruments, one by one, just as the natural scientist had to do—and the formulating of the theories and laws which underlie their use. It is for this reason that I consider the time-span measurement of level of work (described in Chapter 7) and the possible time-frame measurement of level of individual capability (described in Chapter 9) of some interest as possible prototypes. They demonstrate that such derived measurement is at least possible. They suggest that the same search for extensive measuring instruments that proved so important for the development of the modern natural sciences, could be equally important for the human sciences. The search for such instruments is not all that easy. They are not there, like ripe fruit, ready for the picking. But, on the other hand, unless they are specifically sought, as a matter of consciously informed effort, they will never be found at all.

It might, however, still be argued, I suppose, that objective derived measurement of truly human activity is impossible, is a contradiction in terms, for the subjective is not objectively measurable. Thus, for example, it has been asked about the measurement of level of work in time-span of discretion, "Where is the separate objective yardstick? Is not time-span of discretion based simply upon the subjective judgment of the manager about how long it should take his subordinate to complete certain tasks? Surely it is just a matter of subjective judgment, like rating any attributed or preference value. Another manager would have judged differently. It is all individually subjective. There is no objective socially shareable yardstick."

This argument might seem to be justified. There does not at first sight seem to be any materially objective yardstick like a foot-rule or a spectrometer. But there is an objective independent instrument nonetheless. This instrument lies in the objectively definable procedure or operation of getting a manager to make an objectively shareable committed decision about the maximum target completion time he will allow for a given assignment—not his judgment against someone else's judgment, but a *binding decision* in the sense of being objectively, justiciably, binding upon both him and the subordinate by virtue of the employment contract. Decision processes of this kind (as compared, for example, with the more trivial decisions of much research concerned with the throwing of dice), and the resulting objectively stated decisions, are no less independent and objective than is reading a thermometer or a barometer.

I believe that this last point will prove to apply to all derived measurement of properties of psychological and social things. Stated in general terms the proposition is as follows. The measurement and definition of these properties lie in the construction and use of objectively shareable social instruments (social procedures) which commit at least two individuals to joint, or common, or related purposive episodes (defined in terms of time of intention), and stated in such a manner as to be independently justiciable in the case of disagreement either about the original intent or about whether the intent was achieved in the objectively agreed time.

It is this independent open justiciable quality of committed agreement between people that makes nonsense of the idea that social entities are not directly observable like physical entities are. The process and the decision whereby a manager commits a subordinate to a given task in a given time—or to take an example from above, that of an agreement on a term of contract—are just as observable as a granite rock or as a process such as a rainstorm. It requires as part of the observation that we listen to what the people are saying to one another, agreeing with one another. Listening to what they say and agree is what transforms their subjective mental processes into objective facts. And it is precisely this feature—this necessary leaving in of stated perceptions and memories, and desires, and intentions, and social commitment—which makes the most objective possible type of measurement of human activity so human rather than cold and dehumanizing. It is the socially objectified human purpose which allows for the measurement of purpose in terms of the time of intention, and for the measurement of subsequent achievement as against intention, in terms of the time actually taken.

Rating of Individual Preferences Compared with Measurement of Properties

The process of the rating of individual preferences has been completely separated from that of the measurement of objectifiable properties. This separation was made first of all because it was felt to be the most useful, or the most realistic, or, to put it quite simply, the right thing to do. This step, however, poses a serious philosophical problem, a problem which has dogged the whole of our argument about measurement of properties: how can we know whether a phenomenon is an objectively constructed property of a given entity, or a particular person's attitude toward that entity or personal judgment about it? The particular approach adopted to the solution of this philosophical problem (whether or not this approach is made explicit or left implicitly rattling around) exerts a powerful influence upon the development of quantitative methods and concepts. In considering this problem the following formulation suggested itself.

The most commonly held view is that both properties and preferences are quantifiable by a common process called measurement. Underlying this view there are some unstated philosophical assumptions, the most important of which is that properties and preferences are fundamentally the same, because, or at least so it is argued, if you can assign transitively ordered numbering to any phenomena then the phenomena must be of the same kind.[30] It is this view I wish to question.

If we ask, for example, "How long, how valuable, and how beautiful are given entities, and where are they located?" our sense says that length is an objectively existing social construction about the entity beheld in the sense that it can be measured in a way that does not vary between observers, whereas value and beauty will vary with the preferences in the eye of each individual beholder. Preferences are more in the nature of properties of the observer than properties of the entity observed. They are a statement of the attitude toward the entity of each individual person judging it. Preferences by different people with respect to the value or utility of the same thing may range from one extreme to the other: from beautiful to ugly, good to bad, desirable to undesirable, depending upon

[30] N.R.Campbell, for example, considers length, price, and beauty all to be properties of the entity judged. The only distinction he makes is between those properties that can be dealt with by the scientist because they can be subject to agreed quantitative judgment (he puts length and price together), and those which cannot (beauty). As he puts it: "The ... property, beauty, can certainly not be measured, unless we accept the view ... that beauty is determined by the market value .... Beauty is not a property with which science can have anything to do, because no agreement can be obtained for judgments concerning it; it could only become a matter for scientific investigation if a relation can be established between it and some property, such as price, concerning which agreement can be obtained." Op. cit., p. 268.

the observer. As every economist and shopkeeper knows, one man's meat is another man's poison.

There is a difference between phenomena which are possessed by an entity in the sense of an objectively definable quality of it (properties), and those which are related to individually variable attitudes toward the entity. The problem is to define the nature of the existence and the location of these phenomena. The answer lies, I think, in the process of natural concatenation and the meaning of zero. Let me recall. In the case of the measurement of properties, we started with an entity in existence. That existence gave us our fundamental measure of property: the property of spatial length in the case of physical entities and the property of temporal length in the case of psychological and social entities. If we now subtract an amount of length (spatial or temporal, as the case may be) equal to the amount of that property, we get both a zero measurement for the property and an entity which no longer exists. Property and entity disappear together.

Now let us examine the process of rating individual preference-based judgments about an entity. Here I am including values such as goodness, and esthetic judgments such as beauty, utility, and desirability, and the attitudes which lead to preference ratings and to choices. The entities may be material, or psychological, or social. The essence of such preference rating is that it is always individual and always relative, always comparative. There is no absolute scale. One entity is always more or less beautiful than another, more or less valuable, more or less preferred, desirable, good, satisfactory, or useful.[31] This feature leads to a central focus upon relative ordering relationships and ordinal or interval scaling.[32]

Natural concatenation, however, cannot be carried out with such ratings. The reason for this limitation lies in that important number, zero. Zero under conditions of preference rating does not mean no entity. It refers only to a person's attitude toward the entity. The most that it can mean is no value, no interest, no beauty, no utility, unpreferred, a neutral or disinterested attitude as far as any given rater is concerned. It not only does not mean zero entity, it means a definitely existing entity, there to be valued, or rated, but not arousing any single sense of value common to everyone. Indeed, the value can become negative, and rating scales are, of course, frequently constructed in this way: the negative rating,

[31] Ward Edwards emphasizes this point in his writings on multi-attribute utility measurement. See, for example, W. Edwards and M. Guttentag, (1975), "Effective Evaluation," in C. A. Bennett and A. Lumsdaine (Eds.), *Evaluation and Experiment*. New York: Academic Press.

[32] See, for example, L. D. Phillips, (1974), *Bayesian Statistics for Social* . London: Nelson, and New York: Crowell; and D. Lindley, (1971), *Making Decisions*. London: Wiley.

quantified in negative numbers, expressing ugliness, disutility, badness, unsatisfactoriness, rejected rather than preferred.

We can now discover the meaning and location of preferences. Going through the same exercise as for measurement of properties, we start with an entity in existence. We can now rate the entity for particular attributes, by comparing it with other entities of the same type. We can then reduce the rating to zero (or even minus), but the entity does not disappear as it does in the case of entities and properties; it continues to exist as a zero- or minus-rated entity.

In short, an entity can be said to possess properties in the sense that entity and property coexist: zero property—zero entity. By contrast, entities and preferences about them may be said to be coincident but not coexistent: one is not absolutely dependent upon the other. Hence, zero preference—plenty of entity. Properties are measured on absolute-zero-based scales; preferences are rated on relative-zero-based, ordinal, interval, and ratio scales. Rating of preferences, however, can yield only ordinal and interval scales. Such ratings cannot possibly yield ratio scaling, since there is no absolute zero: and if the data can be treated so as to derive a ratio scale, then it is because an objective property has been made explicit. And, be it noted, even the ordinal and interval scales obtained by preference rating are different from those obtained by measurement: they have neither absolute zero nor natural concatenation, at least one of which is present in the case of the ordinal and interval measurement of objective properties.[33]

As against this analysis of a fundamental distinction between properties and attributes of entities, it might, I suppose, be argued that properties are themselves psychological attributions to objects, that they are merely the way human beings see their world. It should be obvious,

[33] A brief study by Robyn Dawes, for example, illustrates the consequence of the failure to distinguish between scaling for the measurement of properties and scaling for the rating of preferences. He took five scales which had been constructed for the rating of various judgments in other studies—for example, ratings of authoritarianism, of the value of a police–community relations program, and of the "attractiveness" of various Army jobs— and modified the questions so that they referred to estimates of the height of a number of people. He then demonstrated that these scales "gave a very reasonsable estimate of height." He does not state, however, that height is an objectively definable property of a physical organism, and not an attributed value, and that the estimated scaling for height is based upon an absolute–zero which starts at the floor upon which the subjects are standing. Estimation of height is an estimate of a measure of a property: the estimates can be compared with a ratio–length measure; and zero height, no person! The values for which these rating scales were originally developed can be used only to compare the distribution and range of preference attributions made by a population of individuals using the scales; and a zero rating expresses only the relative indifference of the individual rater. The fact, therefore, that the scales when adapted for height measurement give reasonable estimates is not relevant with respect to their use for collecting information about the preference judgments which individuals attribute to things. R.M.Dawes, (1977), "Suppose We Measured Height with Rating Scales Instead of Rulers?" in *Appl. Psych. Meas.*, Vol.1, pp.267–273.

however, that I am consistently referring to a phenomenal construction of the world, and in no way suggesting that the operational definition of properties by measurement somehow tells us about the nature of the noumenal world, about what is "really" out there. Entities, properties, and preferences are all aspects of the human construction of the world.[34].

It is essential, however, that we distinguish in that construction between phenomena that are objectively constructed and hence coexistent with particular entities and those which are individual opinions and merely coincident.

Lebesgue and Measurement Mathematics

The foregoing argument receives support from the developments in measurement mathematics associated with Henri Lebesgue.[35] Lebesgue sets out three fundamental conditions for the definition of magnitudes assigned to bodies:[36]

(a)–a magnitude G is said to be defined for the bodies belonging to a given family of bodies if, for each of them and for each position of each of them, a definite positive number can be assigned;

(b)–if a body C is partitioned into a certain number of sub–bodies, C_1, C_2, C_3, ... C_p, and if for these bodies the magnitude G is g on the one hand and g_1, g_2, g_3, ... g_p, on the other, we must have

$$G = g_1 + g_2 + \ldots\ldots + g_p;$$

(c)–the family of bodies for which a magnitude is defined must be sufficiently extensive for every body in the family to be reducible to a single point by successive reductions (without its leaving the family) in such a way that in the course of these reductions the magnitude decreases continuously from its original value to zero.[37]

[34] This fact holds just as true for psychophysical measurement of color, of sones, of lumens, for which precise relationships are obtained between regularities in psychological responses to variations in the quantity of particular properties of a physical stimulus. See S. S. Stevens, (1959), "Measurement, Psychophysics, and Utility," in C.W.Churchman and P.Ratooch (Eds.), *Measurement: Definitions and Theories*. New York: Wiley.

[35] The source is Henri Lebesgue, (1966), "Measure and Magnitude," translated and edited by Kenneth May, and published as one part of *Measure and the Integral*. San Francisco: Holden–Day, Inc.

[36] Lebesgue, op. cit., pp. 128–131.

[37] These propositions have been expressed mathematically in terms of "systematic methods of assigning a 'length' or measure to [subsets of the real line], and any such assignment should 'behave reasonably.' Experience defines this reasonable behaviour as follows:

Let S be a collection of subsets of $[a,b]$ which is closed under countable unions, that is, if

$A_1, A_2, \ldots \in S$, then $\bigcup_{n=1}^{\infty} A_n \in S$.

These conditions clarify the distinction between entities and their objective properties, and separate both from preference judgments about entities. Lebesgue distinguishes between two phenomena: first there are sets or families of bodies, and second there are the measured magnitudes of bodies composing the sets. Bodies are simply physical entities as used by Lebesgue: and he deals with the numbers of bodies in a set by enumeration.[38] Measured magnitude is the phrase used to describe properties: Lebesgue, in his examples, uses length, mass, volume. That he does not concern himself with ratings of preferences can be shown by reference to the conditions.

The first condition would apply both to measurement of properties and to rating of preferences: the duration of a service contract (a measurable property) may be 4 years (say) and its value (a rated attribute) £10,000.

The second condition applies to property measurement, but does not necessarily apply to preference rating. If a 4–year contract is partitioned into a 3–year followed by a 1–year contract, then the partition of its measured magnitude of duration maps on to $4=3+1$. But it does not necessarily follow that the assessed, or rated, £10,000 value will divide into an assessment (rating) of £7,500 for the first part and an assessment of £2,500 for the second: the parts of the divided contract may have enhanced value, or diminished value, or no value at all.

(i) A *set function* on S is a function which assigns to each set $S \in S$ a real number.
(ii) A set function μ on S is called a *measure* if
(a) $0 \leqslant \mu(A) \leqslant b - a$ for every $A \in S$,
(b) $\mu(\phi) = 0$,
(c) whenever $A \subset B$ and $A, B \in S$, then $\mu(A) \leqslant \mu(\beta)$ (monotonicity),
(d) whenever $A = \bigcup_{n=1}^{\infty} A_n$, where $A_n \in S$ for $n = 1,2 \ldots$ and $A_n \cap A_m = \phi$ for $n \neq m$, then

$$\mu(A) = \sum_{n=1}^{\infty} \mu(A_n)$$

(countable additivity).

We will also require that subintervals of [a,b] be in S, and the Lebesgue measure of an interval be equal to its ordinary length. Finally, we hope that every subset of [a,b] will be in S, that is, we will be able to take the measure of any subset of [a,b]. However, we will see that this will be impossible if we wish to retain countable additivity (property (d) in the definition).'' H. J. Wilcox and D. L. Myers, (1978), *Introduction to Lebesgue Integration and Fourier Series*. Huntington, New York: Krieger Pub., pp.15 and 16.

[38] Despite the fact that he defines measurement as all processes of assigning numbers to things, Campbell in practice also distinguishes measurement from enumeration and counting. He uses these terms to refer to the process of determining the numerosity of the members composing a system. These members are described as objects. He ties enumeration and counting to the identification of objects in a very fundamental way: ''The division of our experience into individual objects is simply based on the conditions in which the conditions necessary for enumeration are found. What has been said about the division of a system into members, or individual objects, applies equally to the selection of the operation which is to be used to distinguish the objects in the process of counting.'' N.R.Campbell, *Foundations of Science: The Philosophy of Theory and Experiment*. p. 298.

The third condition also applies to property measurement but does not apply at all to preference ratings. A service contract is readily reducible by successive reductions to a mathematical point, in such a way that its intended length of time reduces continuously to zero; it is so reducible regardless of who does the reduction. But as far as value rating is concerned, a shortened contract may in fact increase in value to the point where completion in no time at all would be infinitely valuable; what actually happens will vary according to the particular person who is doing the valuation (the rating).

It may be noted that it is the combination of the second and the third conditions from which are derived the concatenative process and the absolute–zero base-line which were found to be critical for the quantification of properties, and for equal–ratio scaling. The absence of these conditions characterizes preference ratings, and precludes equal–ratio scaling of preferences.

In short, the Lebesgue conditions for the rigorous mathematical definition of the measurement of magnitudes (the assignment of magnitudes to bodies) apply clearly and systematically to the measurement of objective properties of human entities. These conditions thus define such objective properties. It is equally clear that the conditions do not apply to rating procedures for quantifying preferences attributed by individuals to entities. Such preferences require to be defined by other means— by the definition of rating procedures.

Finally, it may be of interest to note how Lebesgue himself fails to apply his conditions rigorously when he turns briefly to examples using social entities. He states: "The number of magnitudes is vast. As we have seen, it includes numbers associated with geometry and physics and also numbers having to do with economic questions, such as the price of a piece of merchandise, the time necessary to manufacture it, etc".[39]

The time necessary to manufacture a piece of merchandise is certainly a measured magnitude of the process. It will be self-evident that it conforms to all three conditions of measurement. It will be equally evident, however, that the price of a piece of merchandise does not so conform. Price does not necessarily reduce as you reduce a piece of merchandise; different people will react differently to such changes. Indeed, the price may increase with decrease in size if miniaturization is valued. In fact, further consideration suggests that price is itself a social entity associated with, or attaching to, the piece of merchandise, and is not a measured magnitude. As an entity, a price itself has a measurable magnitude, namely, amount of money. As you reduce price by successive

[39] Lebesgue, op. cit., p. 133.

reductions to a point, the money quantity reduces continuously to zero. The recognition of social entities and their measured magnitudes (or objective properties) requires special care.

Lebesgue's correct example— that of time necessary to manufacture a piece of merchandise— illustrates our general theme. This period of time is the measured time–span of achievement for the process. In order to get this measure, however, there must be a starting point as well as a completion point. The starting point is the point of intention, with an equally measurable time of intention. It is this latter measure which is the true beginning of mathematically rigorous measurement of human entities. Human purpose provides the foundation for the scientific study of human activity.

Summary

In this chapter I have tried to demonstrate the importance of keeping purpose and intention in the analysis and measurement of human behavior, in two ways: first, in the clarification of the definition of psychological and social entities as categories of episode; and second, in the separation out of the objectively definable properties of such entities from the attributed preferences of individuals associated with them. The entities are quantifiable by assigning unity to each case and then using counting (enumeration) procedures. Objective properties of entities are definable and measurable by mapping on to an objective time yardstick marked out with a ratio–length scale. Preferences attributed to entities are definable and quantifiable by ordering personal judgments to ordinal or interval scales by rating procedures.

By taking intentional goal-directed activity and episodes as the starting point for the construction of a scientific study of social and psychological phenomena, it is possible to put the human sciences on precisely the same footing with respect to measurement and quantification as the physical sciences. Paradoxically, the more "human," the more full of meaning, intention, desire, need, will, feeling, we keep our psychological and social sciences, the more quantitative and "scientific" they can become, in the sense both of the rigor and of the elegance of measurement.

CHAPTER TWELVE

Dual Perspectives in Temporal Experience

Our inquiry into the nature of time led us to the apparent conflict which exists between two major philosophical views of time. The two views are, on the one hand, that of time as objective or real, discontinuous readings on a clock, atomistic, the time of *chronos*; and, on the other hand, that of time as subjective, continuous, flowing, the unbounded field of durée and *kairos*. These two views have been brought together into a 2-D conception of time. In this chapter I shall seek to transform the apparent inconsistency and conflict between the two viewpoints into a dilemma which grips the two views in a live relationship one with the other. It will give us a conceptual context which will help to make the 2-D view more readily understandable. The handling of the dilemma lies in the recognition of a continuous oscillation in cognitive organization between two polar opposites: a discrete object-dominated organization and a continuous-field organization.

These two poles—discrete object and continuous field—are linked by the experiencing self. They characterize human experience not just of time but in general. They form the two alternating modes of experience, the perspectives from which we view the world. A balanced view calls for a continual oscillation between one perspective and the other. This oscillation gives us in one perspective our world of objects set in a continuous field background. It equally gives us in the other perspective our world of feelings, sensing, intuition, judgments, nous, set against a background of discrete objects which lie ready to jump forward in their turn as we seek to formulate our feelings and sense of things.

It is when the essential process of oscillation between these two perspectives gets stuck, fails to function, that much trouble starts. The discrete object perspective by itself lands us with a mechanistic view of the world and a mechanistic behaviorist view of humanity. The continuous field perspective by itself loses us our ability to discriminate, and ends up in a kind of flowing mysticism. Positivist science and mystical withdrawal are opposite sides of the same coin—a coin whose two sides do not want to know what the other side is doing and are resistant to being turned over.

The chapter ends with a review of the conception of a 5-D (x, y, z, ts, ti) view of the human world, complete with cognitive oscillation between the discrete object and continuous field perspectives. This conception rests upon our 2-D construction of the form of time. It gives a foundation for scientific study and understanding of individual behavior and social institutions which leaves their content of meaning and intentionality—in short, their humanity—intact.

The Duality of the Experience of Time

I find myself, in this last stage of writing this book, with a difficult problem. The problem is one which has trailed along throughout the book. It does not go away however much one might try to overlook it, disregard it, or push it off. It is this. The question of the form of time has raised at every turn the more general question of how we organize experience. In particular, how is the experience of the form of time organized? In general, how is any experience organized?

The problem has shown in our continually being thrown back to the unresolved contradictions between Parmenides, Leucippus and Democritus, and Heraclitus and Bergson, atomism and phenomenology, intuitionism and realism, Being and Becoming. These contradictions have found expression in various pages in a series of antinomies or dichotomous pairs. Here is a list of those I have been driven to use in trying to describe the ways in which we experience time: atomism and field; constancy and flux; objective and subjective; discontinuity and continuity; bounded and unbounded; focused and unfocused; static and dynamic; verbal and nonverbal; bodily and mental; cognitive and conative; reasoned and sensed (impassioned); concrete and abstract; disjunctive (or, v) and conjunctive (and, $\wedge$); lawful and just; logical and rational; context and content; figure and ground; conscious and unconscious (preconscious); points and regions; prescribed limits and discretionary content; succession and intention; earlier–later and past– present–future; clocks and clouds; organized and unorganized; formed and unformed; space-as-place and space-as-force; knowledge and sense; foreground and background. These pairs I have used, and the list can easily be extended.

What is there then about the way we experience things that causes us to think in this two-sided split-minded way not just about time but about nearly everything else, including our scientific and philosophical outlooks? Not only that, but the two sides for some reason or other appeal to the emotions in such a way that each side seems to exclude the other, seems to send it packing. Thus many of the arguments about the nature of time seem to come from the taking up of positions on one or other side of an ill-defined but nonetheless real fence, it being easy enough to find adversaries who will take up the other side. Battle is readily joined between the atomists and fluxists, experimentalists and experientialists, realists and intuitionists, mechanists and mystics, positivists and poets.

In these days, however, when thinking in dichotomies has become more suspect, and pluralistic outlooks more acceptable, it would perhaps seem unnecessary to retain these antinomies as separates. The thing to do would seem to be to treat them as false dichotomies, and to resolve them. But resolve them how? There is more than one way to do so. One might, for example, simply deny the dichotomies, forget about them. Or they might be absorbed into a monist view, formed into modern versions of a uniform atomism or of monads. Or they might be frozen into a Cartesian type of dualism. Or they might be differentiated further and transformed into a pluralistic view of the universe, a view which is recurrently popular.[1] Or they might be retained as pairs of opposites, but brought together as duals, as in dualities in logic or in dialectics, and treated as dilemmas.[2]

My own feelings and intuition push me spontaneously to the last course; not to dualism but to dualities, dialectics, dilemmas. Reflection reinforces this view. Thus, there is no use denying the dichotomies; they have persisted and recurred too powerfully in human thought to go away that easily. Nor does monism or pluralism offer a satisfactory way out. They have in common a suppression, a pushing to one side, a washing over, of the strength of the antinomies. Nor will Cartesian dualism help: that way lies a division of the world itself into material and psychological events which must perforce thereafter run along in parallel, a preservative in which to fix the dichotomies, not to resolve them.

What is called for is not a monist, or dualist, or pluralist view of the world. Indeed it is not a conception of the world itself that is needed, but rather a conception of how the experiencing self creates its world, or how experience itself is organized. Approached from this angle, ex-

[1] Among the most recent writings, for example, Thomas Nagel expresses this outlook in his seeking for pluralist solutions to such philosophical issues as the mind–body problem, subjectivity and objectivity, panpsychism. Thomas Nagel, (1977), *Mortal Questions*.

[2] Language here is tricky, and the philosophical distinction between dualism, on the one hand, and duality and dual, on the other hand, needs noting with care. Dualism refers to the Cartesian view of the world as split irreconcilably between body and mind. Dualities and duals refer to pairs of interconnected and interactive concepts, which may or may not be opposites, such as figure and ground, or the positive and negative poles of a magnet, or the alternation of the truth values in the and/or conjunctions in truth-table logic.

perience—primitive experience, raw experience, that is—seems to organize itself not so much in undifferentiated fields, or in multiplicities, or in parallel dualisms, as in whole ideas divided into paired opposites, that is to say, into dualities.[3]

It is by means of intellectual work that we sort out ideas, carry out analyses and syntheses, think up formulations and hypotheses, and proceed toward the construction of unified field theories or toward the understanding of the complex interaction of multiple variables which influence events. But underneath these mental activities there remains nevertheless the actuality of the experience of interacting dichotomies which continuously suffuses our thought. Because the problems of formulating the nature of time appeared to be so dominated by such antinomies, it seemed to me to be necessary to confront these formidable pairs directly, as properties of experiencing rather than as properties of the world, and to treat them as pairs, as dualities, as opposites, each term of which is symbiotically dependent upon its opposing term for its very life, its very existence.[4]

Alternation of Dualities

How then does experience deal with dualities, with dilemmas? One cannot keep the two sides in focus at the same time, at least not without becoming mentally cross-eyed. The only other possibility would seem to be to alternate between the two sides, but without putting either side right out of the picture. This process of alternation, or of oscillation, is a familiar

[3] Fraser expresses the matter well, in relation to time: "There exists a real duality between knowledge felt and knowledge understood. . . . Their seeming incompatibility may stem from the same source of incomplete perceiving power as does 'the conflict between *lived time* apparently understood and the *idea of time* as an entity which when critically examined is found to be replete with obscurities and unsolved problems.' The idea emerges that C.P. Snow's 'two cultures' are basically not those of the humanities and the sciences, but of knowledge felt and knowledge understood. This division, then, might relate to the duality of the two *times* mentioned above. . . . Communication between the humanities and the sciences, using time as a common theme . . . is important, . . . The total creativity of society may depend upon a harmonious dialogue." J.T. Fraser, *The Voices of Time,* pp. xx and xxi. The quotation within the passage is from the essay by A.C. Benjamin, "Ideas of Time in the History of Philosophy" in the same volume.

[4] I am purposely leaving aside the possible significance of the current work on the split brain by Sperry and others. Their work suggests a neurophysiological basis for the interaction of verbal discrimination of object and touch-and-feel sense in our experience of the world. That picture coincides precisely with my argument. This research, however, is still in too early a stage to draw firm conclusions. And even if it were more advanced, it would not prove conclusive for our purposes. The experience with which we are concerned is a function of the total organism, and not of the brain alone or any part of the brain. The split-brain research may eventually demonstrate that there are neurological subsystems which would make it possible for human beings to behave in the dual manner with which I am concerned. It could never be true, however, that the split-brain organization is the physiological location of mental dualism: it would be a necessary but not a sufficient condition—a necessary subsystem of the physical organism, but not sufficient to explain the functioning of the human system as a whole. R.W. Sperry, (1969), "A Modified Concept of Consciousness."

ordinary feature of everyday life. There is no river without banks, no banks without a river, but in order to perceive both you must oscillate from one to the other; and so between sky and horizon, picture and frame, print and page, self and *umwelt*, and any other figure and ground.

This notion of oscillation seemed on the face of it to cope comfortably with the experiental dichotomies I had encountered: no beads of time without a continuous temporal thread on which to string them; no temporal constancy without flux; no discontinuity without a continuity surround; no temporally bounded zones without an unbounded surround; no focus without an unfocused surround; no verbal communication of time without a nonverbal context; no context without content; and so on. The members of each pair demand each other in order to exist; and all pairs are reversible with respect to which member contains the other. The lesson seemed obvious: keep the pairs as pairs, each side linked to the other, holding the other in its arms, acting as feed to the other, needing the other for its own fulfillment, but each alternating with the other in taking over the front of the stage, making of experience a process of cognitive oscillation from one perspective to the other. Such a cognitive oscillation fits the restless alternation between fluidity and fixity, between continuous flow and discontinuous succession,[5] in the experience of time.

But the problem then arises as to whether each of these dualities of experience is independent of the other, and whether they are equally significant. Is it possible, for example, to be oscillating in perspective between boundedness and unboundedness, atomistic and field view, constancy and flux, focused and unfocused, concrete and abstract, succession and intention, sensing and knowledge, all at the same time? Or independently of each other, each pair oscillating in its own time? Or is there some overriding principle governing and patterning the whole, so that as an oscillation occurs there is a reorganization of all the pairs as though concerted by a conductor in accord with a score? Furthermore, it must be asked whether such oscillations are merely swings back and forth

[5] Some of the difficulties in distinguishing continuity from discontinuity can be seen in Russell's definition of these concepts, namely: "Continuity applies to series (and only to series) whenever these are such that there is a term between any two terms. Whatever is not a series, or a compound of series, or whatever is a series not fulfilling the above conditions, is discontinuous. (The objection to this definition is that it does not give the usual properties of the existence of limits to convergent series which are commonly associated with continuity. Series of the above kind will be called *compact*) . . . the series of rational numbers is continuous . . . the letters of the alphabet are discontinuous." Bertrand Russell, (1903), *The Principles of Mathematics,* p. 193.

Difficulties arise, I believe, because of the attempt to apply the concepts of continuity and discontinuity to mathematical series. There is an inherent contradiction between the concepts of continuity and of series. Russell is defining two types of discontinuous series: finitely divided, and infinitely divisible. Continuous flow, as I am using the concept, has no divisions, other than the cut at each end of a line or episode, which allows it to be dealt with analytically as a whole.

between unchanging opposites, or whether there is some form of movement, of progression, of dialectic, as each swing occurs.

The answers to these last two problems seemed to be self-evident. In the first place, there is a strong coherence, a strong pattern, a strong similarity, between all the terms on one side of the various dichotomies, and between all the terms on the other side. On the left side, as it were, all the terms for discontinuity, discreteness, objectification, materiality, discrimination, in short for objects and atomism, hang together. On the right-hand side are all the terms for continuity, flow, field, subjectification, formlessness, and fluidity. Shift one and you shift them all, for, like oil and water, the terms from the right do not mix with those on the left, and vice versa. They must therefore oscillate together, as two separate internally coherent groups. There is an underlying common binding duality with the formless undiscriminated field on one side and formed discriminated objects on the other.

Then in the second place, it seemed equally evident that since we are dealing with human experience the oscillations must progress, for experience cannot occur without at least a modicum of learning. The duality between formlessness and form is not unconnected with the Platonic ascent from undiscriminated flux to the Forms, and of the dialectical outlook in general, which proved so useful in the theory of levels of abstraction which I described in Chapter 9 and which assumes qualitative changes in levels of abstraction at critical points of change in time-frame.[6]

Out of these considerations the following formulation suggested itself. In particular it is a formulation which precludes any possibility of treating time in the oversimplified terms of flux or durée alone, or of atomistic discontinuity alone, or of A-series or B-series or neither. It may do justice to each.

[6] My colleagues Isaac, O'Connor, and Gibson have utilized such a view in their experimental work and systematic studies of levels of abstraction and discontinuous stages in human development and in propositional logic. Central to their argument is the notion that mental activity is organized in dualities, and that there is always movement from the unformed to the formed, to the unformed at a higher level (or regression to a lower level). They show how the self and *umwelt* develop as the poles of a primitive relationship between self and object. It is this relationship, the experiened need for something, which precipitates the discrimination first of the thing-pole as satisfying the need, and then of the self-pole as the recipient of satisfaction of the need. Experience then swings through the need relationship from pole to pole, with increasing discrimination and development arising from satisfaction, and regression and loss of discrimination resulting from frustration. They posit such a movement to be contained, for example, in the T and F terms of ordinary truth tables: confusion (unform) underlies F, and discrimination (form) underlies T. The movement to higher levels of logical proposition always begins as a movement from F to T. See especially D.J. Isaac and B.M. O'Connor, "A Discontinuity Theory of Psychological Development," and R.O. Gibson and D.J. Isaac, "Truth Tables as a Formal Device in the Analysis of Human Actions"; both in Jaques, Gibson and Isaac, (1978), *Levels of Abstraction in Logic and Human Action*.

Duality of Discriminated Object and Continuous Field

I shall propose that there are two key perspectives within which the human mind constructs its concept of time (and of other experiences as well). These two perspectives are connected; they are the opposite poles of a duality. They comprise the perspective of unformed and continuous field in flux, and the perspective of formed discriminated objects momentarily gripped in unchanging discontinuous state. A persistent difficulty which obstructs our ready perception of this interactive duality is the tendency of our conscious minds to simplify experience by coming down on one side of any duality and disregarding the other, leaving us with only one hand clapping. And the favored side is usually that of discriminated objects rather than continuous field, since objects seem more solid, more stable, more substantial, more real.

I can illustrate this difficulty by the way we tend to see the reversing face-vase figure (Figure 4-1). We tend to report that we see either the faces or the vase; that is to say, we are conscious of whichever is in focus, in the foreground, at the moment. The alternation is thus reported as occurring between two separate things: faces and vase. Whichever is not in focus would seem to have disappeared. In fact, however, the alternation is between two pairs of things, between two dualities. These two pairs are: first, the white vase set in a black (faces?) background; and second, the two black faces separated by a white (vase?) background. No figure, no background; but also, no background, no figure.

The state of the background which to all intents seems to disappear is difficult to describe; it is problematic. It is not quite a pair of faces or a vase, but it is nearly so—it is tip-of-the-tongue so. The background is in a state of consciously available readiness to become foreground. It is what is meant by the consciously available pre-conscious mental function.[7] It is part of all ongoing consciously organized experience. But what is not problematic—and this is the point of my illustration—is that the figure–ground oscillation is an oscillation not just between two figures but between two figure–ground pairs, between two figure–ground dualities.

This point can be illustrated by reference to the face-vase drawing in Chapter 4. With the vase in black and the faces bounded only at the vase but unbounded for the rest, the vase becomes much more sharply delineated and focused, while the faces are small bounded edges of an otherwise unbounded field. The oscillation is now as follows. First, there is a clearly focused object, the vase. It is a discontinuous discrete object, easy to hold as figure. It is seen against an unbounded white background

[7] The Freudian conception, outlined above in Chapter 5.

which contains in a vague way the outline of the two faces. This first figure–ground organization oscillates with the second which is more difficult to hold on to. It is composed of two faces whose heads float off into an unbounded white space. It is possible to be aware of both faces at once but only in a relatively unstable floating way, as very partial boundaries of an otherwise unbounded field in a state of flux. This field grips the discrete vase object which it is very easy to be aware of strongly organized and available in the background.[8]

In summary, the cognitive oscillation which is described is that between two major perspectives as two poles connected and related by the experiencing self: one pole is the perspective of discontinuous, discrete, and discriminable object foreground (which is easy to put into words) with unbounded field background; the second pole is a more weakly organized partially bounded field foreground (which is nearly impossible to put into words) with an organized object background. I shall refer to them briefly as the discriminated-object dominant perspective $[DO]^{cf}$ representing discriminated object foreground (DO) and continuous field background (cf), and the continuous-field dominant perspective $[CF]^{do}$ representing the continuous field foreground (CF) and the discriminated object background (do).

This discriminated-object dominant ~ continuous-field dominant[9] figure–ground oscillation in perspective $[DO]^{cf} \sim [CF]^{do}$ can be applied to our 2-D analysis of time as discriminated and discontinuous atomism on the one hand and as fluid and continuous durée on the other.

The Discriminated-Object Dominant $[DO]^{cf}$ Perspective on Time

The organized experience of the two-dimensionality of time is that of a duality between, on the one hand, knowledge of atomistic discrete time units in the foreground of experience with a background awareness of flux; and, on the other hand, the foreground awareness of flux and durée against a peripheral awareness of discrete time units ticking away in the background of experience.[10] It will be useful to consider each of these perspectives in turn. Each can be abstracted and used on its own for various analytical or scientific purposes. It is important to note, however, that any such abstraction is merely a convenience and not a piece of

[8] This tendency to give superiority to objectification is a common characteristic of physicists and was tellingly criticized by Cassirer. See, for example, the quotation from his *Substance and Function* given in footnote 11, Chapter 2.

[9] I shall use the symbol ~ to represent "oscillating with."

[10] As Kermode has described it, the sequence tic-toc is discriminated foreground, as against the toc-tic sequence which is always formless even when you bring it into the foreground of perception. (Kermode, *The Sense of an Ending*.)

reality. The reality of experience is that of an oscillation between durée punctuated by atomism, and of atomism contained within durée.

First, then, let us consider the discriminated-object dominant [DO]^{cf} perspective on time, and its implications. From this perspective it is the discriminated atomistic view of time which is in focus. This view is the one which brings the temporal axis of succession into the center of the stage. We can dart around from one time recording to another—the time my watch says at the moment; the time the bell rang ten minutes ago; the time of Caesar's death. There are any number of discrete points of time upon which to focus. Each is retrospective; each is objective.

Behind this focused and verbalizable knowledge of discrete times, of discontinuous points in time, there is a vague awareness of flux, of things changing continuously, of unboundedness, of the felt field of past–present–future, of the restless interchange between memory, perception, desire, and intent. This background sense of flux, of the continuous field of temporality, gives the focused times of retrospective earlier–and–later a moving context in which to live and breathe and not to die.

It is this perspective of the axis of discrete times given depth by its right-angled conjunction with the axis of continuous time, which coincides with one particular side or pole of the experiential antinomies, some of which I mentioned above. The terms of this polarity are all members of the serial atomistic picture of the world. This perspective gives us our experience of the concrete organized thing, discriminated and distinct, things which can stand still or which can move, but even when they move they continue to be identifiable things moving from one discrete place to another. Such motion is concrete in the sense of the motion of solid pieces of concrete and not the all-mixed-up field type of motion of cement, sand, and water slewing about in a mixer before it has set into identifiable objects.

In terms of mental processes, this world of thingumification is of conscious knowledge, of logico-critical precision, of Dedekind-cut arithmetic. Physics starts here—with early stages characterized by precise measurement; then there follows the equal precision in experimentation with experimental and control groups objectively identified and bounded and separated from each other. And in this century we have the development of the Bohr conception of the atom, and the notion that change occurs in discrete and constant quantities.

From the [DO]^{cf} perspective, single factors can be isolated and studied in their own right. Prediction can be handled on the useable assumption of discontinuous cause-and-effect sequences. It need not matter in practice that causes cannot be got into continuous conjunction with effects, since this is the mentally constructed world of discontinuity, of discreteness. So long as we do not confuse ourselves into believing that it is the only perspective, the only real picture of the world, and that it is the object separated from its field which represents somehow the "real" world, the world of real organized objects, an accurate reflection of what the noumenal world is really like, then the notion of the discontinuous causal series can remain the useful tool of research and understanding it has always been.[11]

With respect to the problem of time, it is the discriminated-object dominant $[DO]^{cf}$ perspective within which the conception of time as earlier and later is gripped. It is the time of discontinuity, the B-series, clock time, real time, objective time, the time of discrete points, time which can be cut into lengths, time which can be told, the stateable times which can be put into words and about which we speak. In short, it is this perspective which yields the temporal axis of succession.

It may at first be thought that the discriminated-object dominant $[DO]^{cf}$ perspective is inevitably static, inevitably 4-dimension bound, inevitably the handmaiden of the narrow-minded and rigid kind of positivist empiricism to which more dynamically oriented, more human, more skeptical philosophies object. But this view is not necessarily true. So long as the continuous field (cf) background is kept in the picture, is held continuously in peripheral awareness, then a balanced view of the world—including the 4-D material world—can be maintained, with the availability of oscillation between the two major perspectives as appropriate. This view is the one adopted by any good scientist—adopted by him in action, that is, in his actual work—but not necessarily adopted by him in his formulations of what he supposes his scientific method to be.

Given this moving balance between two perspectives, the objectifiable concreteness of the discriminated-object dominant $[DO]^{cf}$ perspective can be used to good effect in the human sciences. It gives the formulated legal context, for example, to encompass and set limits to legal judgment; it gives the temporal axis of succession along which purposeful episodes (and the 4-D context of history) can be recorded; it

[11] The solution to Hume's uncertainty about cause–effect sequences because the final conjunction could never be established, is resolved, I believe, once we note that discontinuous cause and effect is only our $[DO]^{cf}$ abstraction which forms the foreground of the continuous field background. It is in our background awareness that we experience the continuous flow of the fields of force which give the necessary causal flow and not simply the sufficient cause– effect jumps of the discontinuous foreground figure. When we assume the $[CF]^{do}$ perspective, our analysis of causation can switch from sequential cause-followed-by-effect, to an analysis in terms of the dynamics of the changes inherent within a field of force at a given point in time. Both constructions of causality are necessary.

organizes the self as object, thus establishing the notion of the discrete individual; it makes possible the construction of separate entities and the counting of significant social units and thereby the field of social statistics; it perceives the prescribed limits of all action; it is essential for all logic—traditional, sentential, propositional. And withal, it makes it possible to isolate and explicate the structure of social and psychological processes, to abstract them, and to put them into descriptive or quantitative language; whatever might be the artificiality and the limitations of this conventional process of abstraction, it lies at the core of human verbal interaction and civilization.

The Continuous-Field Dominant $[CF]^{do}$ *Perspective on Time*

The $[CF]^{do}$ perspective on time is the dual of the $[DO]^{cf}$ perspective. It can be briefly elaborated by reviewing the previous section and by putting the opposite statement in each case. This perspective is that of the moving field, of unbounded continuity, of clouds rather than clocks, of the flow of the mixture of cement, sand, and water, of rivers, of electromagnetism, of waves. It gives the human world of social interaction and of self-as-agent, of syntax rather than grammar, of conation, feeling, passion, intuition, of nonverbal communication and communion, of belief, of losing oneself in action, of creativity, of spontaneity and responsibility, of values, of discretion, of judgments within the law, of sensed goals and intentions, of Becoming, of responsiveness to poetry and music and other evocative experience.

With respect to the problem of time, it is the continuous-field dominant $[CF]^{do}$ perspective which gives the conception of time as durée, as continuity. It is the time of simultaneity, of past–present–future, the time of the field of interactive memory–perception–desire–purpose, of the A-series, of continuous change and flow, time without length and without discrete points, the time which can be felt but which by itself cannot be told, the time without end and the time of sensed eternity. In short, it is this perspective which yields the temporal axis of intention.

This world is often thought of as necessarily unscientific, the world of the second culture. That view is not correct. Neither of these two perspectives is either more scientific or more artistic than the other. Both perspectives, oscillating, interpenetrating, are necessary for both science and art, necessary for the scientist's intuitive sense of the world, and for the artist's concreteness of use of paint, or stone, or tones, or words.

As in the case of atomism or object-dominance, given the moving balance between the two major perspectives, the sensed purposeful feel of the continuous-field dominant $[CF]^{do}$ perspective can be used to good effect in the human sciences. It does justice to the flow of human episode within the boundaries of beginnings and endings either planned or actual; it allows for the interpretation of this flow in history as much as in psychoanalysis; it gives the setting for force-field analysis, whether the 4-D force-fields of electricity, magnetism, and light, or the 5-D force-fields of human behavior; it separates out discretion, intention, and prediction, and gives the needed perspective for their analysis. And withal, it lays the foundation for the natural science type of objective equal–ratio–length–scale measurement of the inherent properties of intrapsychic processes, social interaction, and social institutions, so that the psychological and the social sciences, and artistic and intuitive life, can counterpoint and illuminate each other, rather than contradicting and negating each other.

The Experiencing Self

I have treated the discriminated object and continuous field perspectives as two poles of a relationship, symbolized as $[DO]^{cf} \sim [CF]^{do}$. The relationship between these two poles is that of the activity of the experiencing self, of the person. The experiencing self does not exist in isolation: it exists, and can only exist, in our continuous interaction with other people and with things. It is out of this activity that the experiencing self generates its discrimination of object out of undiscriminated field, and then its awareness of field itself. We gain in knowledge and skill from this oscillation between growing discrimination of objects and increasing confidence in functioning in the continuous field mode necessary for action.

It must be noted, however, that our reflexive awareness of the experiencing self does not remain unaffected by the cognitive oscillation described. In the $[DO]^{cf}$ discriminated-object dominant perspective the self also is perceived as object. We are aware of the limits of our body image as somehow the objective representative or manifestation of ourselves. We can observe ourselves in action, seeking things, doing things, saying things, accomplishing things and, above all, meaning things.

It is in this mode also that we know ourselves as facts, know of our identity. We do so on the temporal axis of succession. We identify ourselves as the same person at this time (the retrospective reconstruction of a moment ago) as the person retrospectively known at an earlier time, both these times being linked within the context of the background awareness of memories and feelings which are the cf context of the $[DO]^{cf}$ perspective.

In the $[CF]^{do}$ continuous field perspective the self is no longer discriminated, but is rather the subject of a vague awareness as being in continuous motion, in action, in flux. This awareness is of the self-as-agent. It is the state of mind in which we give ourselves up to action, lose ourselves in what we are doing, relinquish our need for the moment to have clear-cut boundaries to grasp and to clutch, and get on with what needs to be done.

In this mode we are aware of ourselves as continuants, not by retrospective construction of ourselves at discrete points in time but by the direct awareness of ourselves in the process of continuing existence itself. Such awareness could not occur were we totally lost in action or in thought. It is requisite that we retain our grip on the structure of our discriminated world via our awareness of this structure in the background (the do context of $[CF]^{do}$), a background to which we can flick our attention from time to time in maintaining our personal orientation.

Awareness of self-as-object and self-as-agent is not, then, necessarily a matter of different selves in the sense of personal disconnection. It is rather a matter of alternating perspectives in the construction of a complex self, and a complex of selves, growing, developing, changing through time, as our experience drives forward, at times gaining in discrimination, and at times also regressing and losing in richness of self-awareness and self-knowledge. This process of development requires that the discriminated object perspective and the continuous field perspective be held tightly as one pair of connected and related poles which I have expressed as $[DO]^{cf} \sim [CF]^{do}$. If that connection is lost or ruptured, then a troubled construction of our world will occur. Let us consider some of the consequences of such a condition.

Rupturing the Object-Field $\{[DO]^{cf} \sim [CF]^{do}\}$ *Duality: Some Consequences*

It is possible, and indeed commonplace, not only for individuals but also for philosophical and scientific systems to get fixed upon one or other pole of the duality of discriminated object and continuous field. They do so by cleaving the duality, rupturing it, splitting it, so that effective oscillation cannot take place. This splitting and polarization leaves the person ensnared either in a static lifeless world or in an uncontrollably fluid and confusing one.[12] Both states are dis-

[12] Crites links his analysis of the narrative temporal quality of experience to a unified conception of self—a self, however, with two strategies, abstraction and flow. He then argues that the splitting of these two strategies between body and mind has had grave consequences. As he puts it: "The power to abstract makes explanation, manipulation, control, possible. On the other hand we seek relief and release in the capacity to contract the flow of time, to dwell in feeling and sensation, in taste, in touch, in the delicious sexual viscosities. . . . But the modern world has seen these two strategies played off ever more violently against one another. One could show how the reification of mind and body has killed modern metaphysics by leading it into arid controversies among dualistic, materialistic, and idealistic theories. But this comparatively harmless wrangle among post-Cartesian metaphysicians is only a symptom of the modern bifurcation of experience.

fjturbing, if not to the individual then to his relationships with others.[13]

Let me first consider the consequences of fixation upon discriminated objects (the [DO]^{cf} perspective) at the expense of the continuous field [CF]^{do} perspective, and in extreme cases even with the cf context denied. Such a fixation leads not just to a materialist outlook but to a mechanistic materialism which circumscribes and limits the conception of reality, the conception of what really matters, to what can be seen, touched, organized. It is the outlook of that type of logician who believes that all truth is the truth of logical deduction, and that ethics, morals, values, intuition are to be treated as second-class citizens, if not to be ruled out altogether. It is expressed in the narrow experimental positivist view that all that can be said to exist is what can be demonstrated by controlled experimentation, and that what cannot be so demonstrated is not respectable knowledge. Any idea that there might be an unconscious level of mental functioning is simply not acceptable.

Along with this kind of rigid empiricist materialism which sees the world exclusively in terms of discrete atoms and objects (knocked about like billiard balls as Gregory Bateson puts it), there goes a tendency toward an exclusively serial outlook. One thing leads to another. For every effect there must be a unique cause; or if there is a complex of causes, it is referred to as overdetermination. Behavioristic psychology, positivist social and economic research and engineering, simplistic cost-benefit approaches to social policy and to life are the net result, with a tendency to explore and manipulate people and to use them as objects and as instruments.

In severe form, in psychological disturbance, this atomistic fixation on object dominance shows in extremes of concretism in outlook and in expression. Symbolism and metaphor present difficulties, everything

Its more sinister expression is practical: the entrapment of educated subcultures in their own abstract constructions, and the violent reaction against this entrapment, a reaction that takes the form of an equally encapsulating constriction of experience into those warm, dark, humid immediacies. One thinks of Faust in his study where everything is so dry that a spark would produce an explosion, and then Faust slavering and mucking about on the brocken. Against the inhumanly dry and abstract habitations of the spirit that have been erected by technological reason, the cry goes up, born of desperation, to drop out and sink into the warm stream of immediacy. Within the university the reaction and counter-reaction have been especially violent in the humanities.'' Stephen Crites, (1971), ''The Narrative Quality of Experience.''

[13] This process of mental splitting is central to the psychoanalytical theories of Melanie Klein. It is a process characteristic of the very earliest infant stage of development (which she refers to as the paranoid-schizoid position) in which splitting the world of objects into good and bad is the prime defense against anxiety. These splitting processes continue throughout life in the primitive layers of the mind, with more or less disruptive effects depending upon the extent to which they have been worked through or left unresolved. One of their disruptive consequences is expressed in intellectual rigidity and inhibition and in confusion, which are outstanding features of precisely the kind of splitting and rupturing of the field of experience which I am describing. See Melanie Klein, (1975), *Collected Writings*.

tending to be taken literally. Hanna Segal has described how such people use symbolic equations rather than symbolism, the difference being that in symbolic equation the symbol is taken to be the thing itself, the word is the thing.[14] This kind of concreteness exists to some extent in all of us. The issue is to what extent, and how much of it is split-off. Too much concreteness, too much split-off can disturb the free oscillation between object dominance and field dominance with consequent restriction of outlook, or frank emotional disturbance.

To return to the problem of time, there are many examples of what happens to our conceptualization of time under conditions of frozen polarization and fixation on to the atomism discrete-object perspective. There is the narrow static atomic conception of Democritus and Leucippus, the restriction to a notion that the only "real" time is clock time. There is a resulting reification in which it is taken for granted that time flows, the problem being to discover whether it is unidirectional, or bidirectional, or reversible, and in what direction it is going. And there is the inability to cope objectively with intentionality.

I think it is because of the sense that that lopsided one-sided perspective is wrong in some unrecognized way that the paradoxes of Zeno have remained so intriguing for two thousand years. This wrong perspective is that of pure atomism, pure [DO], without any acceptance whatever of field and flow. Thus, if like Zeno one assumes that the world is made up only of discrete and discontinuous objects, and that time also is totally discontinuous, then, as Zeno rightly thought, the only possible conclusion is that there can be no such thing as relative motion. In such a world, arrows, Achilles, tortoises, soldiers, at a point in time in space, cannot really move continuously toward a moving target. They are limited to jumping from one point to the next, in a discontinuous series of infinitesimally and decreasingly small jumps. There is no flow in this world—the world in the atomistically split-off mental perspective of Zeno, that is.

In the world of process observed as such, in which the cognitive mode of the unbounded field of continuous flow interacts with the point-at-a-distance discriminated object mode, we can obtain the advantage of seeing arrows reach the target, of seeing Achilles overtake the tortoise, while at the same time we can abstract points in the process and calculate velocities and acceleration. In thus abstracting, we do not have to lose all awareness of flow in the split mechanistic materialism which is the source of the paradox. Thus it is that what appears to be a paradox is in fact the reflection of a splitting of mental process.

[14] H. Segal, (1957), "Notes on Symbol Formation."

At the other pole we have $[CF]^{do}$ without $[DO]^{cf}$, and we get all the signs and symptoms of fixation upon fluxion and continuous fields without the leavening influence of the organizing force of the discrimination of discrete objects and without the retrospective analysis of momentarily stilled structure. It is a shift into a timeless world of flux and movement, a world of endless unverbalized contemplation and inaction. The continual exploration of the sources of intention and desire, the supposed sources of life, is substituted for the expression of intention in action.

With respect to the problem of time, this split-off continuous-field perspective leads to the rejection of the temporal axis of succession, as Heidegger did explicitly. Bergson did not do so, but unfortunately gave the impression that he did, in his great emphasis upon durée.[15] What we are left with is a disembodied psychological time, a purely phenomenological view of time, a view which readily enough deteriorates into transcendental mystical retreat.[16]

When this polarized perspective is not only deprived of alternation with its polar opposite but also subjected to denial even of its discriminated object context (that is to say, [CF] alone rather than $[CF]^{do}$), then the perspective on time deteriorates into a frankly mystical view of the world. All is eternity, or even beyond eternity. There is insufficient form and discrimination even to give the context for time-frame activities. Confusion and withdrawal, with only the minimum activity necessary for the maintenance of life, become the order of things.

A less extreme form of this kind of self-indulgent mysticism shows in what Christopher Lasch has called the culture of narcissism—a culture of indulgent sensuousness which avoids feelings by cultism, emotional shallowness, and new consciousness movements.[17] He describes, in another study, the narcissistic withdrawal which characterizes the split sink-

[15] His reply to those critics who accused him of overlooking the world of discriminated reality is trenchantly given in *The Creative Mind*. He writes, for example: "The philosopher who upholds the determinism can plead his cause nonchalantly . . . simple, clear and right. He is easily and naturally so, having only to collect thought ready to hand and phrases ready-made: science, language, common sense, the whole of intelligence is at his disposal. Criticism of an intuitive philosophy is so easy and so certain to be well received that it will always tempt the beginner. Regret may come later, unless of course . . . there is personal resentment toward all that is properly spent. That can happen, for philosophy too has its Scribes and Pharisees." (P.37). Then later he argues for both science and metaphysics: "To sum it all up, what is wanted is a difference in method between metaphysics and science: I do not acknowledge a difference in value between the two." (P.42). Henri Bergson, (1965), *The Creative Mind*.

[16] How easy it is nearly to fall into this outlook shows in Norman Brown's struggle to formulate unconscious processes. "The doctrine of the unconscious," he wrote, "properly understood, is a doctrine of the falseness of all words, taken literally, at their face value, at the level of conscious. . . . The true meanings of words are bodily meanings . . . the unspoken meanings. What is always speaking silently is the body. . . . To recover the world of silence, of symbolism, is to recover the human body." Norman O. Brown, (1966), *Love's Body*, pp. 257 and 265.

[17] Christopher Lasch, (1978), *The Culture Of Narcissism*.

ing into the inner world of the unbounded temporal mode: "The contemporary cult of sensuality implies a repudiation of sensuality in all but its most primitive forms. . . . Ideologies of impulse gratification and pleasure seeking gain the ascendancy at the very moment that pleasure loses its savor. A narcissistic withdrawal of interest from the external world underlies both the demand for immediate gratification . . . and the intolerable anxiety that continually frustrates this demand."[18] This sinking into an uncritical restless continual fluxion into the new becomes a negation of time flux, for the underlying aspiration to circumvent old age and death is too readily apparent.

In its most extreme form the retreat into unbounded time (and space) is the psychotic retreat into a world of confusion, of wordless nothingness, and total loss of competence in social interaction.

Dual Perspective and 5(3+2)-D Human World

A balanced formulation of the human world, the social world, demands both a comfortable oscillation between the two major perspectives and a 5-D viewpoint. Indeed, it is only with a 5-D viewpoint that an effective oscillation in perspective on human affairs is possible. An exclusively 4-D construction of the living world limits us to one temporal dimension only. If we choose the axis of succession (ts), as we do for the natural sciences, that tends to push us in the direction of the mechanistic outlook which accompanies excessive favoring of the discriminated-object perspective with which temporal succession is mainly operative. Alternatively, if we choose the axis of intention (ti) on its own, then that tends to push us in the other direction; namely, toward the mystical outlook which accompanies excessive favoring of the continuous-field perspective. I have already elaborated upon the consequences of any tendency to fix upon either perspective in our formulated constructions of time and of the nature of our world.

Our 2-D construction of temporal axes of succession and of intention can support a more realistic construction of the human world than can a unidimensional conception. This construction can encompass both the discriminating definition necessary for composing the structure of psychological and social things and contexts, and the continuous-field nondiscrete functions necessary for identifying the processual contents of social purposive episodes.

Effectively balanced oscillation in perspective between the poles of the object-field duality is not simply there for the asking. It must be worked for. In particular it is the field-dominant perspective which can

[18] Christopher Lasch, (1977), *Haven in a Heartless World*, p. 183.

be lost sight of. The object-dominant perspective is strong, it pushes itself forward, like the vase in Figure 4.1. It is the unbounded continuous field which is more difficult to keep hold of. Effort and toil are called for if the two poles of the objective-field duality are to be "rubbed against one another," as the forms of understanding which Plato described in the *Seventh Letter* (344B) need to be rubbed together for the true path of knowledge to be followed as described in the *Phaedo* (79D) "through anguish and resistance . . . and the soul liberated from aimless wandering."

In ordinary circumstances, and at its most effective, the oscillation between the two modes is rapid and unnoticed. Psychological confidence in the oscillation—that is to say, confidence that one will not suddenly become split—allows us to lose ourselves in whatever we are doing, to lose ourselves in living, to take part fully in human experience, to live life to the full.

In doing any kind of work, for example, it is necessary to be able to lose oneself in the work. A pianist, for instance, must be able to lose himself in performance. But he cannot *completely* lose himself, or he is lost. To lose himself without becoming lost means to be able to allow his performance to flow while still keeping contact in preconscious awareness with where he is and with what he is doing. In allowing his performance to flow he must retain control and maintain his context, but without being disturbed by this awareness, without being aware either of context or of flow. It is a state of suspended awareness of flow, of context, all in one, the state of being lost in doing something. It is a state that is perhaps more dramatically expressed in the performance of the members of a great string quartet.

This losing oneself in doing things applies to any activity whatever. All work is genuinely creative in this sense. There is no such thing as uncreative work. What we tend, however, to call creative work, is work of very long time-span; unfortunately it leads us, incorrectly, to consider any other work, by contrast, as not creative.[19]

This sense of being able to lose oneself in an activity, of being able to rely upon the tiny-amplitude fibrillating unnoticed miniscule moments of differentiated cognitive organization, is of course not achieved without effort, without learning, without practice. During practice we tend to lean heavily upon the discriminated-object orientation, breaking processes up into discretely manageable bits, and being painfully consciously aware

[19] People who are widely regarded as geniuses work, I believe, in time–spans of 50 years and more. Such time–spans give them a personal time–frame which runs beyond their adult lifetime—and everyone else's as well. This feature gives the timeless and mysterious quality to their work, and means that there is an important sense in which their work is always incomplete.

of what we are doing and of how long it is taking us to do it. The art of art, then, is to get it all to flow together.

Judges and Juries

Another way in which the necessary flowing together of object and field perspectives in a 5-D framework can be usefully illustrated is in the duality of judge and jury. I choose this illustration because the various elements can be seen clearly because of their differentiated institutional forms. Thus, in discussing what he deems to be the proper relationship between jury in criminal cases and judge in criminal appeals in English law, Lord Devlin states: "On demeanour the jury is final. What an appellate court does is to stand back and to look at the case immobilised on paper. Before the jury the case is unwound as on a revolving cylinder; before the judges it is spread out. The jury sees the whole performance, but cannot except in memory move from one incident to another; the judges can freeze the significant points, study, and correlate them. It is the examination of the case on paper that gives to appellate judges the detachment which is the essential quality of a court of review."[20]

In a telling reinforcement of the same argument about the solidity of law as discriminated context as against the fluidity of discretion within the law, he writes:[21] "Those who live under the rule of law have as their entitlement the right to look first to the law for their protection and to the *discretion* of a judge only supplementarily *in those recesses of justice into which the law is too solid to flow*." (Italics mine.)

Then, felicitously calling attention to a famous passage of the historian Gibbon that "the discretion of the judge is the first engine of tyranny,"[22] Lord Devlin goes on to restate his belief "in the jury and in case law for the same reason, that they are both restraints on judicial supremacy. I see the value of discretionary power, but where it is to be used widely, I would rather it was spread upon the diversity of the jury than concentrated in judges sitting singly or in like-minded groups; and where it is to be given to the judiciary I would rather it was used in accordance with the corporate wisdom and accounted for accordingly, than left to the individual judge to say yea or nay."

Both judge and jury move in an explicitly 5-D world: indeed the accurate determination both of intention and of the actual time of events is one of the main objects of the legal process. But judges both in trials and in appeals have to adopt the atomistic viewpoint. Their vantage point is the case statically fixed, analyzable, focused, knowable, and relatable

[20] Patrick Devlin, (1979), *The Judge*, p. 170.
[21] Op. cit., p. 201.
[22] Gibbon, (1982), *The Decline and Fall of the Roman Empire*, Vol. V, p. 328.

to the atomistic focused formulated legal context, with formulated reasons stated to support and explain their decision, thereby adding to the corpus of law. For the jury the task is the opposite. It is for them to follow the trial, get the feel of events, experience it in terms of their mental fields of memory, perception, and expectation, and to use their discretion as best they can to bring forward a fully made decision for reasons which need no explanation or analytic justification. Their decision contributes to the community flow of feelings of justice but not one iota to the law.

In the interplay of the duality of judge and jury lies justice for the individual and legal orderliness for society. Without both poles of this duality justice is not simply incomplete, only half there: justice is transformed into something else. Judge without jury is tyranny; jury without judge is lawlessness, the crowd which condemned Socrates to death. Judge and jury each without the other, would be a product of a 4-D world. With each other, interactively, they are an expression of a 5-D world view, with interacting intentionality and succession.

History and the Balancing of Memory, Perception, and Intent

Santayana wrote that a person who does not remember his past is doomed to relive that past. That statement is true: but it is incomplete. We might add: A person who does not perceive his present is seriously out of contact; and a person who does not have desires for his future, who is devoid of expectation and anticipation and intention, is doomed to an arid and empty life. Equally we might state the converse of these truisms: A person who knows only his memories is doomed to imprisonment in his past; a person who perceives only the present is cut off from the richness of his experience and from concern about the future and the excitement of intention; and a person who lives only in and for the future will be lost in fantasy and imagination and will achieve little.

In short, any fragmenting and unbalancing of the free interplay and interweaving of memory, perception, and will and desire, as a unified field, will lead to an impoverishment and disruption of the psychic life. The steady modification and development of memory, perception, and will and desire, each by the others and by itself, on the axis of intention, and sufficient awareness of these changes to date them on the axis of succession, are essential elements in a reasonable and reasoned existence. These elements are dependent upon the functioning of the object-field duality.

If, however, the past is truly lived and experienced only in the present, in the field context of perception, desire, and intent, what are we to make of history? Do events really exist in the past? Can they be accurately reconstructed? Can we, as the positivist historians who follow

Comte would suggest, really know what happened? Can even one person's history be accurately reconstructed, as in psychoanalysis?[23] Or is it all a matter of interpretation, a matter of the idiosyncracies of each historian or each psychoanalyst and his data, and the social context within which he makes his interpretation?

The central questions are how to avoid the mechanistic positivist historicism of Mommsen, Foustel de Coulanges, Comte, and Taine,[24] which seeks to establish what "actually happened" and to explain it in physical cause and effect terms, while at the same time not being thrown into a romantic historicism which seeks to find some sort of purpose in the movement of history, a universal purpose superior to the purposes of individuals. I believe that such questions can be to some extent clarified by adopting a 4-D approach to the history of the physical world and a 5-D approach, with oscillating perspectives, to living history, along the following lines.

The 4-D reconstruction is sufficient for establishing the history of the physical world and its physical objects. We start with artefacts and detritus currently in existence. They can be placed in three spatial dimensions. They can then be dated in one temporal dimension—on the axis of succession—by carbon testing, by counting rings on trees or by inspecting teeth, by proximity to other things found in a particular geological stratum which in turn has been dated.

In the same vein, a limited 4-D analysis might serve to determine constructions of human history in a restricted positivist sense; that is to say, to determine whether some objectively perceivable event occurred or not, and if so, when it occurred. Such, for example, are the dates of Caesar's invasion of Britain, the existence of the Greek academics, Socrates's death by hemlock, Neanderthal man, Henry VIII's wives, the French Revolution, World War I. Here we have artefacts and written documentation, or even eyewitness data, to organize in 4-D, including the temporal axis of succession.

There is no obvious advantage, however, in restricting human historical analysis to such a 4-D frame of reference. More can be accomplished, and with equal or even additional rigor, by adopting a 5-D outlook. This outlook allows, in the first place, for an objective dating of history retrospectively along the temporal axis of succession, without

[23] Sigmund Freud, (1937), "Constructions in Analysis," in *Standard Edition,* 1961, Vol. XXIII, p. 257.

[24] Gertrude Lenzer (Ed.), (1975), *Auguste Comte and Modern Positivism,* describes the relationship between Comte's positivist historiography and the mechanistic social engineering by means of which he proposed to advise rulers on how to order their states. She shows how social scientists like Durkheim and Skinner follow this tradition.

having to go all the way, for example, with Ranke of whose rigorous objectivism Cassirer states: "[His] sole purpose was to show 'how it actually happened,' and he would have preferred 'to blot out his own self likewise,' so to speak, in order to let only the historical events and mighty forces of the centuries be heard."[25]

It is important, however, not to confuse retrospective dating of events on the temporal axis of succession with something called the real, or actual, past. A retrospective dating construction is simply what it is; namely, a present-day construction from presently existing data. An interesting construction no doubt, but only the palest verbal shadow of what the actual past event must have been like, as experienced by those involved at the time.[26]

Moreover, such a retrospective dating construction does not put any meaning into the events concerned. To do that requires an act of historical interpretation—an interpretation, or assessment, or judgment, of what the events might have meant to those who lived through them. This analysis requires the attribution of meanings, purposes, goals, intentions, to people in history, and to the picture of their social interactions and institutions—a difficult task at best, and one which can never be completely achieved, since no records whatsoever can exist which can give objective evidence of the unconscious feelings and aspirations of those people and of the meaning of their interrelations, since even they themselves would have been only partially aware of them.

Nevertheless, constructions can be made from records of stated purposes, and from interpretations from other data. Be it noted, however, that these purposes are concerned with individuals and their interactions, and not with grand notions about the purposeful movement of society, or the grand purposes of history. They are in fact a 5-D filling-out of history, using the temporal axis of intention to describe the sense of past, present, and future of people at particular historical points on the axis of succession. It is this 2-D analysis of time in history that can sometimes lead to extraordinary discoveries such as Von Schliemann's Troy, when to the high degree of probability usually inherent in axis of succession dating is added the usually lower probability analysis of the dominant individual purposes of key figures in a society.

A complete social history, then, is a 5-D history, oscillating between the [DO]^{cf} perspective of retrospective dating of events on the axis of

[25] E. Cassirer, (1950), *The Problem of Knowledge*, p. 224.

[26] As Riegel puts it: "Our inability to learn 'how it really was in history' should disturb us as little as our failure, according to Kant, to recognize 'the thing as such.'" K.F. Riegel, (1978), "A Dialectical Interpretation ofTime and Change," in Gorman and Wessman (Eds.), *The Personal Experience of Time*, p. 94.

succession, and the $[CF]^{do}$ perspective of interpretive construction of the purposes of people and the timing of those purposes on the axis of intention. The combination of these dimensions, and oscillation between their different perspectives, gives history with meaning, even though the meaning needs continual rethinking and revision in the light of the dominant outlooks of each succeeding generation. The interpretation of its historically recorded past, and the seeking of new records about the past, are part of the current ideology of any society. This interpretation and ideology are changing events in the space–time manifold just as are all other events. That is how history remains a living subject: historical truth versus historical distortion is not a question of objective fact versus falsehood—it is a question of the intentions of the historian, of intentional seeking for truthful interpretation versus the intention to distort.

The same 5-D considerations apply to psychoanalytical reconstruction. There is the recovery of lost memories of childhood, and the filling in of the sequence of significant events and their dates. Then there is the buildup of the meaning of those events to the patient, and that is a quite different matter. For the events exist for the patient as memories, as past in the present. It is that past in present memory, that "vast and infinite interior space,"[27] which can be directly interpreted in terms of the patient's present state of mind, in which memory, desire, and intent express the interrelated past, present, and future.

The significance of history too lies in its present meaning for people, and not in what in any absolute sense, or even highly probable sense, might have occurred at an earlier time or times. I use the term "earlier" advisedly, for the past of history lies not in a past, but in the present interpretation of where we have come from, in the interpretation of earlier times which constitutes the present sense of pastness which is such an important feature of any people's awareness and picture of itself.

Words and Meanings

Finally, language, too, depends upon the double cognitive $[DO]^{cf} \sim [CF]^{do}$ organization. The point-at-a-distance organization allows for the construction of discreteness of words. The unbounded-field organization allows for the fullness of meaning inherent in the intentions lying behind the choice and use of discrete words. In the two orientations are contained both the explicit point-at-a-distance thing-references of words, and the metaphoric symbolic contents, the meanings and feelings behind the words as things, the depths of language, the human, poetic, feeling, and revealing content, accompanied by the nonverbal expressive communi-

[27] St. Augustine, *Confessions,* X; VII.

cations in tone, timing, movement, and gesture, which may either reinforce or contradict the explicit verbal contents, but in either case will extend their meaning. Meaning is thus a continuously unfolding process, into which we dip from time to time to extract circumscribed and focused quantities of consciously articulated verbalization.

The idea that verbalization itself might emerge from the oscillating interaction between two modes of cognitive organization is a proposition which it might be worthwhile to record for future consideration. In the unbounded-field [CF]^{do} perspective the preconscious-unconscious mode of cognitive organization is dominant. Verbal language lapses, sinks into the preconscious ground, ready to become explicit again. Is this situation, then, the one in which the intrapsychic field of force is such as to cause movement to take place across the unconscious-preconscious boundary, so that previously unconscious, unverbalized, field-organized mental processes take on the outlines, at least of discrete entities? If so, they would at this stage become available for verbalization during the switch to the discriminated object, [DO]^{cf}, point-at-a-distance mode when consciously focused figures interact with preconscious ground.

If ordinary language is a discriminated object [DO]^{cf} phenomenon backed by the continuous field mode, then scientific formulation and measurement and quantification are also dependent upon the [DO]^{cf} perspective which they express as a special type of rigorously defined language. I have tried to show, however, that effective quantification requires the oscillation between the two cognitive modes. By means of the continuous-field [CF]^{do} mode, we feel ourselves into the phenomena we study, getting our sense of what goes where, what is linked to what, and what is worth formulating and measuring.

Out of this sense of things we isolate the locale of things of significance for study, of processes worthy of observation, of absent things and relationships whose existence we intuitively feel and strive to get hold of. Then by psychological effort we maneuver these flux-sensed impressions bit by bit into the discriminated object [DO]^{cf} mould, squeezing and shaping and naming them, so that they become explicit and recognizable new items, new creations called discoveries—such as electricity, or penicillin, or chromosomes.

The trouble starts, however, if we try to treat the fiction as reality in connection with human processes and episodes. Human life is a continual dilemma: it must hold the balance, the duality, between the spoken and the unspeakable, between knowledge and judgment, between theory and flux, between Being and Becoming. However, this duality, this dilemma, is more than a matter of tidiness and symmetry. It is a matter of life and death for the individual and for society.

Industrial society, with its industrialized mind and its emphasis and reliance upon the natural sciences and technology, has overvalued and been grossly overimpressed by the critical, the conscious, the verbal, the brain (especially the cortex), the mechanical robots, technology, everything to do with knowledge, the passing of exams as well as formal qualifications, quizz games, the emphasis on numbers. It has lost its ability sufficiently to value and to feel secure in relying upon the other side of the human equation—the side that contains intuition, judgment, flowing unverbalized sense, the feel of the situation, the deeper sense of simply understanding what is right and wrong and fair and just because it feels right or wrong or fair or just, the sense of the reasonable, the ability to sit back and reflect and remember and to feel a part of one's past and present, and to identify with other human beings, to feel empathy and sensitivity without either false sentimentality or embarrassment.

We need to keep strong contact with our intuitive sense of movement of episodes in an unbounded field—for without that contact we lose our sense of purpose and we lose our sense of contact with the intentions and the purposes, the desires and the will, which make human episodes human. To be able to keep such contact is precisely what Keats has called negative capability; this is, "when a man is capable of being in uncertainties, mysteries, doubts, without any irritable reaching after facts and reasons."

I have tried to show that so long as we do not lose our firm foothold in the unbounded-field sense of human activity, we can abstract episodes by abstracting chunks of process along the temporal axis of succession, not only without losing the life of human existence but enhancing our perception of it. It is for this last reason that an understanding of the form of time is of peculiar importance for the understanding of human life, intellectual and impassioned, knowing and feeling, thinking and acting, resting and working, desiring and willing, reflecting and exploring, lacking and seeking, waking and dreaming, withdrawing and relating—a time for living and a time for dying. Just as a 4-D physical science can leave the world of things intact, so a 5-D temporalized human science may give us the powerful assistance of a scientific understanding, but leave intact our human sense of the human events which that scientific understanding is meant to illuminate.

BIBLIOGRAPHY

Aristotle. 1928. "De Categoriae"; and "De Interpretatione," in W.D. Ross (Ed.), *The Works of Aristotle*. Oxford: Oxford University Press.

Barker, T., Dembo, T., and Lewin, K. 1941. "Frustration and Regression." *University of Iowa Studies in Child Welfare,* 18, No. 1, 1–43.

Barr, James. 1969. *Biblical Words for Time*. London: S.C.M. Press, 2nd Edition.

Bateson, Gregory. 1979. *Mind and Nature: A Necessary Unity*. London: Wildwood House.

Benjamin, A.C. 1966. "Ideas of Time in the History of Philosophy," in J.T. Fraser (Ed.), *The Voices of Time*. New York: George Braziller.

Bergson, Henri. 1910. *Time and Free Will*. Tr. by R.L. Pogson. London: George Allen & Unwin.

———. 1911. *Matter and Memory*. London: George Allen and Unwin.

———. 1965. *The Creative Mind*. New Jersey: Littlefield, Adams.

———. 1965. *Duration and Simultaneity*. Tr. by L. Jacobson. New York: Bobbs-Merrill.

Birdwhistell, Ray L. 1973. *Kinesics and Context*. Harmondsworth: Penguin Books.

Boltzmann, L. 1964. *Lectures on Gas Theory*. Tr. by S.G. Brush. University of California Press.

Bonaparte, Marie. 1940. "Time and the Unconscious." *Int. J. Psychoan.*

Braithwaite, R.B. 1928. "Time and Change." *Proceedings of the Aristotelian Society,* Supp. Vol. 8, 169.

Broad, C.D. 1938. *An Examination of McTaggart's Philosophy*. Cambridge: Cambridge University Press.

Brown, Norman. 1966. *Love's Body*. New York: Random House.

Brown, Wilfred. 1971. *Organization*. London and Exeter, New Hampshire: Heinemann Educational Books Ltd.

———. 1973. *The Earnings Conflict*. London and Exeter, New Hampshire: Heinemann Educational Books Ltd.

Brown, Wilfred, and Jaques, Elliott. 1964. *Product Analysis Pricing*. London and Exeter, New Hampshire: Heinemann Educational Books Ltd.

Butler, R.J. 1955. "Aristotle's Sea Fight and Three-Valued Logic." *Philosophical Review,* 64.

Campbell, N.R. 1920. *Foundations of Science: The Philosophy of Theory and Experiment* (formerly titled *Physics, The Elements*). New York: Dover Publications, 1957.

Cassirer, Ernst. 1923. *Substance and Function, and Einstein's Theory of Relativity*. New York: Dover Books, 1953.

———. 1950. *The Problem of Knowledge*. New Haven and London: Yale University Press.

Coombs, C., Dawes, R., and Tversky, A. 1970. *Mathematical Psychology*. Englewood Cliffs, New Jersey: Prentice-Hall.

Crites, Stephen. 1971. "The Narrative Quality of Experience." *J. Am. Acad. Relig.*, Vol. 39, No. 3.

Cullmann, Oscar. 1964. *Christ and Time*. Tr. Floyd V. Filson. Philadelphia: The Westminster Press.

Danto, A.C. 1976. *Sartre*. London: Fontana/Collins.

Dawes, R.M. 1977. "Suppose We Measured Height with Rating Scales instead of Rulers?" *App. Psych. Meas.*, Vol. 1.

Dennes, W.R. 1935. *The Problem of Time*. University of California Publications in Philosophy, 18.

Devlin, Patrick. 1979. *The Judge*. Oxford: Oxford University Press.

Driesch, H. 1933. *Philosophische Gegenwartsfragen*. Leipzig.

Dummett, Michael. 1954. "Can an Effect Precede its Cause?" *Proceedings of the Aristotelean Society,* Suppl. Vol. 28.

———. 1964. "Bringing about the Past." *Philosophical Review* 73.

Edwards, W., and Guttentag, M. 1975. "Effective Evaluation," in C.A. Bennett and A. Lumsdaine (Eds.), *Evaluation and Experiment*. New York: Academic Press.

Einstein, Albert. 1949. "Autobiographical Notes," in Paul Schilpp (Ed.), *Albert Einstein, Philosopher-Scientist*. La Salle, Ill.: Open Court, 1970.

Eliot, T.S. 1944. "Burnt Norton," in *Four Quartets*. London: Faber and Faber.

Evans, John S. 1979. *The Management of Human Capacity*. Bradford: MCM Publications.

Farlow, J.K. 1959. "Sea Fights Without Tears," *Analysis,* Vol. 19.

Flew, A.G.N. 1959. "Hobbes and the Sea-Fight," *Grad. Rev. Phil.,* Vol. 2.

Fogel, A. 1977. "Temporal Organisation in Mother-Infant Face-to-Face Interaction." in H. R. Schaffer (Ed.), *Mother-Infant Interaction* London: Academic Press.

Fraisse, P. 1964. *The Psychology of Time*. Tr. by J. Leith. London: Eyre and Spottiswoode.

Fraser, J.T. (Ed.). 1966. *The Voices of Time*. New York: George Braziller.

Freud, Sigmund. 1923. *The Ego and the Id*. In Standard Edition, 1961. London: Hogarth Press.

———. 1937. "Constructions in Analysis," in Standard Edition, 1961. London: Hogarth Press.

Gale, Richard M. 1967. *The Language of Time*. London: Routledge and Kegan Paul.

———. 1968. "Some Questions about Time" in Richard M. Gale, (Ed.), *The Philosophy of Time*. London: Macmillan.

Gibbon, Edward. 1862. *The Decline and Fall of the Roman Empire*. London: John Murray.

Gibson, R.O., and Isaac, D.J. 1978. "Truth Tables as a Formal Device in the

Analysis of Human Actions,'' in Jaques, Gibson and Isaac, *Levels of Abstraction in Logic and Human Action*. London and Exeter, New Hampshire: Heinemann Educational Books Ltd.
Goldstein, Kurt, and Scheerer, G. 1939. *The Organism*. New York: American Book Co.
Grant, C.K. 1957. ''Certainty, Necessity, and Aristotle's Sea-Battle,'' *Mind* 64.
Grünbaum, Adolf. 1964. ''Carnap's Views on the Foundations of Geometry,'' in P.A. Schilpp (Ed.), *The Philosophy of Rudolf Carnap*. La Salle, Ill.: Open Court.
Guyau, J.M. 1890. *La Genèse de l'Idée du Temps*. Paris: Alcan. 2nd Edition, 1902.
Hall, E.W. 1959. *The Silent Language*. New York: Doubleday.
———. 1969. *The Hidden Dimension*. London: Bodley Head.
Hampshire, Stuart. 1965. *Thought and Action*. London: Chatto and Windus.
Hays, W.L. 1967. *Quantification in Psychology*. California: Brooks/Cole.
Hegel, G.W.F. 1807. *The Phenomenology of Mind*. Tr. J.B. Baillie. London: George Allen and Unwin; New York: Macmillan, 1931.
Heidegger, Martin. 1962. *Being and Time*. Tr. by J. Macquarrie and E. Robinson. New York: Harper and Row.
———. 1972. *On Time and Being*. Tr. by Joan Stambaugh. New York: Harper & Row.
Hinde, R.A., and Herrmann, J. 1977. ''Frequencies, Durations, Derived Measures and Their Correlations in Studying Dyadic and Triadic Relationships,'' in H.R. Schaffer (Ed.), *Studies in Mother-Infant Interaction*. London: Academic Press.
Hintikka, J. 1964. ''The Once and Future Sea Fight,'' *Philosophical Review* 73.
Homa, Edna. 1967. ''The Inter-relationship among Payment and Capacity.'' Unpublished doctoral thesis, Harvard Business School.
Humphreys, P.C., and Wishuda, A. 1979. ''Multi-Attribute Utility Decomposition,'' *Tech. Report,* 79-2, Brunel University Decision Analysis Unit.
Hunter, W.S. 1913. ''Delayed Reactions in Animals and Children,'' *Behav. Monogr*. 2, No. 1.
Isaac, D.J., and O'Connor, B.M. 1978. ''A Discontinuity Theory of Psychological Development,'' in Jaques, Gibson and Isaac, *Levels of Abstraction in Logic and Human Action*. London and Exeter, New Hampshire: Heinemann Educational Books Ltd.
James, William. 1890. *Principles of Psychology*. New York: Henry Holt & Co.
Jaques, Elliott. 1956. *Measurement of Responsibility*. London: Heinemann Educational Books Ltd.
———. 1961. *Equitable Payment*. London and Exeter, New Hampshire: Heinemann Educational Books Ltd.
———. 1964. *Time-Span Handbook*. London: Heinemann Educational Books Ltd., and Carbondale, Ill.: University of Southern Illinois Press.
———. 1970. ''Death and the Mid-Life Crisis,'' in *Work, Creativity and Social*

Justice. London: Heinemann Educational Books Ltd., New York: International Universities Press.

———. 1976. *A General Theory of Bureaucracy*. London and Exeter, New Hampshire: Heinemann Educational Books Ltd.

——— (Ed.). 1978. *Health Services*. London and Exeter, New Hampshire: Heinemann Educational Books Ltd.

Jaques, Elliott, Gibson, R.O., and Isaac, D.J. 1978. *Levels of Abstraction in Logic and Human Action*. London and Exeter, New Hampshire: Heinemann Educational Books Ltd.

Kelly, G.A. 1963. *A Theory of Personality*. New York: Norton.

Kermode, Frank. 1967. *The Sense of An Ending*. Oxford: Oxford University Press.

Klein, Melanie. 1975. *Narrative of a Child Analysis;* and *Envy and Gratitude;* in *Collected Writings*. London: Hogarth Press.

Koffka, Kurt. 1928. *The Growth of the Mind*. New York: Harcourt.

———. 1935. *Principles of Gestalt Psychology*. New York: Harcourt.

Kohler, Wolfgang. 1929. *Gestalt Psychology*. New York: Liveright.

Krantz, D., Luce, R., Suppes, P., and Tversky, A. 1971. *Foundations of Measurement: Vol. 1, Additive and Polynomial Representations*. New York and London: Academic Press.

Krimpas, George. 1975. *Labour Input and the Theory of the Labour Market*. London: Duckworth.

Lamb, Warren. 1965. *Posture and Gesture*. London: Gerald Duckworth & Co. Ltd.

Langer, Susanne K. 1967. *Mind: An Essay on Human Feeling*. Baltimore: Johns Hopkins University Press.

Laplace, M. le Marquis de. 1820. "Introduction à la Théorie Analytique des Probabilités," in *Oeuvres Complètes de Laplace,* Vol. 7. Paris: Gauthiers-Villars et Fils, 1886.

Lasch, Christopher. 1977. *Haven in a Heartless World*. New York: Basic Books.

———. 1978. *The Culture of Narcissism*. New York: Norton.

Lebesgue, Henri. 1966. "Measure and Magnitude," in *Measure and the Integral*. Tr. by Kenneth May. San Francisco: Holden-Day Inc.

Lennep, D.J. van. 1968. "The Forgotten Time in Applied Psychology." University of Utrecht.

Lenzer, Gertrude (Ed.). 1975. *Auguste Comte and Positivism*. New York: Harper & Row.

Lewin, Kurt. 1935. "The Conflict Between Aristotelean and Galilean Modes of Thought," in *A Dynamic Theory of Personality*. London and New York: McGraw-Hill.

———. 1935. *A Dynamic Theory of Personality*. London and New York: McGraw-Hill.

———. 1942. "Time Perspective and Morale," in *Resolving Social Conflicts*. New York: Harper & Row, 1948.

———. 1952. *Field Theory in Social Science*. London: Tavistock Publications Ltd.

Lindley, D. 1971. *Making Decisions*. London: Wiley.

Lloyd, H. Allan. 1966. "Timekeepers—an Historical Sketch" in J.T. Fraser (Ed.). *The Voices of Time*. New York: George Braziller.

Lucas, J.R. 1973. *A Treatise on Time and Space*. London: Methuen & Co.

McClelland, D.C., Atkinson, J.W., Clark, R.A., and Lowell, E.L. 1953. *The Achievement Motive*. New York: Appleton-Century-Crofts.

MacMurray, John. 1957. *The Self as Agent*. London: Faber and Faber.

———. 1959. *Persons in Relation*. London: Faber and Faber.

McTaggart, J.M.E. 1927. *The Nature of Existence*. Cambridge: Cambridge University Press.

Malrieu, P.H. 1953. *Les Origines de la Conscience du Temps*. Paris: Presses Universitaires de France.

Marsh, J. 1952. *The Fulness of Time*. London: Nisbet & Co.

Mead, George Herbert. 1932. *The Philosophy of the Present*. Chicago University Press.

———. 1934. *Mind, Self and Society*. Chicago University Press.

Mink, Louis O. 1960. "Time, McTaggart and Pickwickian Language," *Philosophical Quarterly*, Vol. X.

Minkowski, E. 1908. Address delivered at 80th Assembly of German Natural Scientists and Physicians, Cologne, September 21st 1908. In Einstein and Others, *The Principle of Relativity*. Tr. by W. Perrett and G.B. Jeffrey, with notes by A. Sommerfield. New York: Dover Publications, 1923.

Murray, H.A. 1938. *Explorations in Personality*. Cambridge, Mass.: Harvard University Press.

Nagel, Thomas. 1977. *Mortal Questions*. Cambridge University Press.

Ovsiankina, M. 1928. "Die Wiederaufnahme von unterbrochenen Handlungen," *Psychol. Forsch.* 11, 302–389.

Pask, Gordon. 1969. "Strategy, Competence and Conversation as Determinants of Learning," *Programmed Learning*, October 1969.

Pepper, Stephen. 1970. *The Sources of Value*. Berkeley, Los Angeles, London: University of California Press.

Phillips, L.D. 1974. *Bayesian Statistics for Social Scientists*. London: Nelson; and New York: Crowell.

———. 1980. "Generation Theory," *Working Paper* 80-1, Brunel University Decision Analysis Unit.

Piaget, Jean. 1937. *La Construction du Réel Chez l'Enfant*. Neuchatel, Paris: Delachaux et Niéstle.

———. 1946. *Le Developpement de la Notion de Temps Chez l'Enfant*. Paris: Presses Universitaires de France.

Pieron, H. 1923. "Les Problèmes Psychophysiologiques de la Perception du Temps," *Année Psychologique*, 24.

Polanyi, M. 1966. *The Tacit Dimension*. New York: Doubleday.

Popper, Karl. 1956. "The Arrow of Time," *Nature*, 177.

———. 1974. *The Philosophy of Karl Popper*. (Paul S. Schilpp, Ed.). La Salle, Ill.: Open Court.

———. 1976. *Unended Quest*. London: Fontana/Collins.

Poullion, Jean. 1966. "Time and Destiny in Faulkner." Reprinted in Robert Penn Warren (Ed.), *Faulkner*. Englewood Cliffs, New Jersey: Prentice-Hall.

Quine, Willard Van Orman. 1953. "Mr. Strawson on Logical Theory," *Mind*, 62.

———. 1960. *Word and Object*. Cambridge, Mass.: M.I.T. Press.

Ramsden, Pamela. 1973. *Top Team Planning*. London: Cassell/Associated Business Programmes Ltd.

Reichenbach, Hans. 1956. *The Direction of Time*. Berkeley: University of California Press.

Richardson, Roy. 1971. *Fair Pay and Work*. London: Heinemann Educational Books Ltd., and University of Southern Illinois Press.

Riezel, K.F. 1978. "A Dialectical Interpretation of Time and Change," in Gorman and Wessman (Eds.), *The Personal Experience of Time*. New York: Plenum Press.

Roberts, F. 1979. *Measurement Theory*. Vol. 4 of Encyclopedia of Mathematics and Its Applications. Reading, Mass.: Addison-Wesley.

Robinson, J.A.T. 1950. *In the End, God . . . .* London: James Clarke & Co.

Russell, Bertrand. 1903. *The Principles of Mathematics*. Cambridge: Cambridge University Press.

Ryle, Gilbert. 1966. "Epistemology," in J.P. Urmson (Ed.), *The Concise Dictionary of Western Philosophy and Philosophers*. London: Hutchinson.

Saint Augustine. 1961. *Confessions*. Tr. by R.S. Pine-Coffin. Harmondsworth: Penguin Books.

Santayana, G. 1937. *Realms of Being*, in *Works*, Vol. 14. New York: Charles Scribner's Sons, Triton Edition.

Sartre, Jean-Paul. 1956. *Being and Nothingness*. Tr. by H. Barnes. New York: Philosophical Library; and London: Methuen & Co.

———. 1966. "On 'The Sound and the Fury': Time in the Work of Faulkner," in Robert Penn Warren (Ed.), *Faulkner*. Englewood Cliffs, New Jersey: Prentice-Hall.

Schachtel, Ernest. 1963. *Metamorphosis*. London: Routledge and Kegan Paul.

Schaffer, H.R. 1977. *Studies in Mother-Infant Interaction*. London and New York: Academic Press.

Schilpp, Paul (Ed.). 1964. *The Philosophy of Rudolf Carnap*. La Salle, Ill.: Open Court.

———. 1970. *Albert Einstein: Philosopher-Scientist*. La Salle, Ill.: Open Court.

———. 1974. *The Philosophy of Karl Popper*. La Salle, Ill.: Open Court.

Schopenhauer, Arthur. 1966. *The World as Will and Representation*. Tr. by E.F.J. Payne. New York: Dover Publications, Inc.

Segal, Hanna. 1957. "Notes on Symbol Formation," *Int. J. Psych-An.*

Skalet, M. 1930–31. ''The Significance of Delayed Reactions in Young Children,'' *Comp. Psychol. Monogr*, 7, No. 4.

Smart, J.J.C. 1963. *Philosophy and Scientific Realism*. London: Routledge and Kegan Paul.

———. 1964. ''Questions about Time,'' in J.J.C. Smart (ed.), *Problems of Space and Time*. New York: Macmillan.

Social Services Organisation Research Unit, Brunel University. 1974. *Social Services Departments: Developing Patterns of Work and Organisation:* London: Heinemann Educational Books Ltd.

Sperry, R.W. 1969. ''A Modified Concept of Consciousness,'' *Psychological Review,* 76, 532–36.

Stamp, Gillian. 1978. ''Assessment of Individual Capacity,'' in Jaques, Gibson and Isaac, *Levels of Abstraction in Logic and Human Action*. London and Exeter, New Hampshire: Heinemann Educational Books Ltd.

Stearns, I. 1950. in *Review of Metaphysics,* 5 198.

Stebbing, L.S. 1936. ''Some Ambiguities in Discussions Concerning Time,'' in R. Klibansky and H.J. Paton (Eds.), *Philosophy and History*. Oxford: Clarendon Press.

Stevens, S.S. 1951. ''Mathematics, Measurement and Psychophysics,'' in Stevens (Ed.), *Handbook of Experimental Psychology*. New York and London: J. Wiley and Sons, Inc.

———. 1959. ''Measurement, Psychophysics, and Utility,'' in C.W. Churchman and P. Ratooch (Eds.), *Measurement: Definitions and Theories*. New York: J. Wiley & Sons, Inc.

Strang, C. 1960. ''Aristotle and the Sea Battle,'' *Mind,* Vol. 69.

Taylor, R. 1955. ''Spatial and Temporal Analogies and the Concept of Identity,'' *Journal of Philosophy,* 52.

Tillich, P. 1936. *The Interpretation of History*. Tr. by N.A. Tasetski and E.L. Talmey. New York and London: C. Scribner's Sons.

Tolman, E.C. 1967. *Purposive Behaviour in Animals and Men*. New York: Appleton-Century.

Trevarthen, C. 1977. ''Descriptive Analyses of Infant Communicative Behaviour,'' in H.R. Schaffer (Ed.), *Studies in Mother-Infant Interaction*. London: Academic Press.

Warren, Robert Penn, (Ed.) 1966. *Faulkner*. Englewood Cliffs, New Jersey: Prentice-Hall.

Weyl, H. 1949. *Philosophy of Mathematics and Natural Science*. Princeton, N.J.: Princeton University Press.

Whitehead, A.N. 1920. *Concept of Nature*. London: Cambridge University Press.

———. 1923. ''The Problem of Simultaneity,'' *Aristotelian Society,* Suppl. Vol. 3.

———. 1938. *Science and the Modern World*. Harmondsworth: Penguin.

Whitrow, G.J. 1975. *The Nature of Time*. Harmondsworth: Penguin.

Wilcox, H.J., and Myers, D.L. 1978. *Introduction to Lebesgue Integration and Fourier Series*. Huntington, New York: Krieger Publ.

Williams, D.C. 1951. "The Myth of Passage," in *Journal of Philosophy* 48, and in his *Principles of Empirical Realism*. Springfield, Ill.: Charles C. Thomas, 1966.

———. 1951. "The Sea Fight Tomorrow?" in P. Henle (Ed.), *Structure, Method and Meaning*. New York: Liberal Arts Press: and in Williams, *Principles of Empirical Realism*. Springfield Ill., Charles C. Thomas.

Von Uexküll, Jacob. 1962. *Biology*. Tr. D.R. Mackinnon. London: Routledge and Kegan Paul.

Zeigarnik, B. 1927. "Uber das behalten von unerledigten Handlungen," *Psychol. Forsch*. 9, 1–85.

Zuckerkandl, Victor. 1956. *Sound and Symbol: Music and the External World*. Tr. by Willard R. Trask. New York: Pantheon Books, Inc., Bollingen Series.

Index